COMMON AND CONTESTED GROUND:
A HUMAN AND ENVIRONMENTAL HISTORY OF THE
NORTHWESTERN PLAINS

In *Common and Contested Ground,* Theodore Binnema provides a sweeping and innovative interpretation of the history of the northwestern plains and its peoples from prehistoric times to the Lewis and Clark Expedition.

The real history of the northwestern plains between A.D. 200 and 1806 was far more complex, nuanced, and paradoxical than often imagined. Drawn by vast herds of buffalo and abundant resources, aboriginal peoples, fur traders, and settlers moved across the region establishing intricate patterns of trade, diplomacy, and warfare. In the process, the northwestern plains became a common and contested ground.

Drawing on a wide range of sources, Binnema examines the impact of technology on the peoples of the plains, beginning with the bow and arrow and continuing through the arrival of the horse, European weapons, Old World diseases, and Euroamerican traders. His focus on the environment and its effect on patterns of behaviour and settlement brings a unique perspective to the history of the region.

THEODORE BINNEMA is an assistant professor in the Department of History at the University of Northern British Columbia.

Common and Contested Ground

A Human and Environmental History of the Northwestern Plains

Theodore Binnema

University of Toronto Press: Toronto

Printed in Canada by University of Toronto Press 2004

ISBN 0-8020-8694-2

Printed on acid-free paper

National Library of Canada Cataloguing in Publication

Binnema, Theodore, 1963–
Common and contested ground : a human and environmental history of the northwestern plains / Theodore Binnema.

Includes bibliographical references and index.
ISBN 0-8020-8694-2

1. Indians of North America – Great Plains – History. 2. Indians of North America – Commerce – Great Plains – History. 3. Fur trade – Great Plains – History. 4. Great Plains – Ethnic relations. I. Title.

E78.G73B56 2004 978.004'97 C2004-901318-1

Cover illustration: *When the Blackfeet Hunt* by John Innes (photography Robert Barrow). Hudson's Bay Company Museum Collection, courtesy of the Manitoba Museum.

University of Toronto Press acknowledges the financial assistance of the Office of the Vice President Research, University of Northern British Columbia.

University of Toronto Press acknowledges the financial assistance to its publishing program of the Canada Council for the Arts and the Ontario Arts Council.

University of Toronto Press acknowledges the financial support for its publishing activities of the Government of Canada through the Book Publishing Industry Development Program (BPIDP).

To my mother and my father

Contents

ILLUSTRATIONS

All illustrations are by the author unless otherwise indicated.

Preface

The roots of this book lie in my master's thesis, a detailed study of Siksika life between 1794 and 1815. When I first confronted the primary documents, two things surprised me. I was struck by how poorly the primary documents addressed the main scholarly themes and debates in North American Indian historiography today. I was also unprepared for the abundant evidence relevant to neglected or downplayed aspects of history. For example, the documents showed that intraethnic and interethnic connections among Indian communities and between Indian and non-Indian communities were far more important than I expected after reading the secondary literature. Documents made it possible for me to understand the roles played by specific people and intraethnic communities and by specific events in the history of the Siksikas. I became convinced that twentieth-century scholarship had neglected some of the most fascinating evidence relating to the history of the northwestern plains. When I embarked upon my doctoral research (from which this book is derived) I believed that I could contribute much more to our collective knowledge by moving in promising new directions than by addressing the debates in the mainstream literature. My previous study of the primary evidence, my judgment of what secondary literature was particularly innovative, and my personal experiences and beliefs influenced the shape of this book from its inception.

Most importantly, this book reflects the intraethnicity and interethnicity that the documents continually alluded to. It is not the history of a certain ethnic group. Although the historical actors clearly identified themselves as Blackfoot, Cree, Assiniboine, Crow, Shoshoni, or Flathead, ethnicity was only one way in which people identified themselves—and the evidence gave me no reason to believe that it was usually the most important. We lack general histories of several ethnic groups on the northwestern plains, but we do not lack tribal histories in general. I intended from the beginning to explore North American Indian history in a way that would recognize the reality of intraethnic communities and interethnic connections. The historical evidence gradually led me to focus on the northwestern plains as defined here, and the scientific evidence confirmed my decision.

The primary evidence gave me no reason to believe that the historical actors were concerned about cultural change and continuity. Indigenous communities did not merely "adjust" or "adapt" to change; they assumed change and often embraced it. For many years most scholars presumed that cultural change implied disintegration. This is no longer the case. Since the mainstream works in the history of the northwestern plains have focused on the themes of cultural change, however, I chose to explore other themes suggested by the evidence. The documents revealed a great deal about the means used by autonomous Indian communities, and by newcomers, to secure their lives. Trade, diplomacy, and warfare were clearly central to northwestern plains history. Given the growing interest in Indian self-government in North America today, a better understanding of the ways in which communities governed themselves and interacted in the past is certainly relevant.

This book is not intended to be a study of the history of North American Indians in the fur trade or of Indian-newcomer relations. Since the 1970s we have learned a great deal about Indian-Euroamerican trade and about bilateral relations between certain ethnic groups and newcomers. Especially in Canada, scholars have long described the complexity of what is now known as the "middle ground" between Indians and Euroamericans, particularly the "middle peoples" that emerged when they interacted. I have interpreted these bilateral relations in the context of a web of relations among indigenous groups. I have also tried to explain the significance of particular leaders, communities, and turning points in the history of the northwestern plains, to recapture some of the contingency of the past and discover the human beings behind the impersonal "clash of cultures."

This book is an exploration of alternatives to the culturalist preoccupations that marked much of twentieth-century scholarship. If it challenges readers to consider how their "Indian of imagination" differs from the "Native American of actual existence," it will have achieved its main goal. I understand the process of historical research and interpretation well enough to realize that this book does not represent a definitive and objective reconstruction of the past. Every work of scholarship is, in part, autobiography. I hope, however, to advance our understanding of an elusive past.

I have been greatly enriched as I carried out this study. My research continually filled me with wonder and admiration for the people who inhabited the northwestern plains over the centuries. I decided to study the history of the northwestern plains because I came to love the region. I have grown to love it more as my understanding of it has increased. I have also benefited from the assistance of many people over the years. My citations express many of my debts, but others need to be amplified. I especially recognize the aid of John E. Foster, under whose supervision this project began. John Foster and I spent many hours in his office discussing and debating issues surrounding my research. I remember those times with fondness. I regret that he did not live long enough to see the completed project. I also thank Arthur J. Ray and Rod Macleod for assuming the supervision of this project after John Foster's death. Ironically, because of the tragedy of his death, I have had the good fortune of being able to work very closely with two of the most gifted scholars of Indian-newcomer relations in North America.

Many others have been exceptionally gracious. Archaeologists at the Alberta Provincial Museum of Alberta have been very helpful. Jack Ives is foremost among them. Others include Rod Vickers, Jack Brink, and Alwynne Beaudoin. Our discussions, including our disagreements, have been very intellectually stimulating. I have benefited immeasurably from our exchanges.

Several scientists have been as selfless as these archaeologists. Thanks to people like A. W. Bailey, Robert Hudson, and Hal Reynolds, my efforts to understand the grasses and bison of the northwestern plains were pleasant and rewarding.

Many people have contributed in other important ways. I thank Bob Irwin, Gerhard Ens, Carolee Pollock, Paul and Jennifer Hackett, and Willy Dobak.

Two other groups of people deserve thanks: the archivists at the Hudson's Bay Company Archives (HBCA) and the staff at the University of Alberta's Interlibrary Loans Office (ILLO). The documents of the HBCA are a much-underrated international treasure. The employees at those archives bend over backward to make the treasure accessible. The staff at the ILLO responded with good humor to my incessant requests for microfilm from the HBCA.

Ultimately, my own family has invested and sacrificed the most so that this project could go ahead. I thank Helen, Derek, Kathryn, and Josiah for their patience and understanding.

EDITORIAL PROCEDURES

Passages cited from other sources are quoted exactly. Errors and idiosyncrasies in spelling, punctuation, grammar, and capitalization are common in primary documents and appear in this work uncorrected, without the distracting use of "[*sic*]." Raised letters at the end of abbreviations in the original document (for example, "rec'd") are lowered to the line. Occasionally I have included clarifications in square brackets in quotations.

TERMS USED FOR NORTH AMERICAN INDIAN GROUPS

There is no consensus in Canada or the United States about how best to refer to indigenous North Americans. In fact terms in common use in one country are sometimes unacceptable or unfamiliar to many people in the other. Whenever possible I have attempted to refer to specific groups; but in other cases I have followed the University of Oklahoma Press policy of referring to indigenous North Americans as "American Indians" or "Indians" or have used terms such as "indigenous people" or "aboriginal people." I have also used the latter two terms when referring to groups that may include certain mixed-blood communities (metis or proto-metis) that were separate from any Indian community. Similarly I use "Euroamerican" to refer to all residents of North America of European extraction.

The terms used in this study to refer to specific American Indian ethnic groups are typically those that appear frequently in primary documents and are commonly understood among English-speakers today. Although neither this usage nor any other is fully satisfying, there are several reasons for this approach. It makes the study more accessible to many potential readers. Most American Indians today employ a prevalent exonym (name by which they are known to outsiders) when they identify themselves in English, just as those who consider themselves *français* identify themselves as "French" when speaking in English. In many cases a similar exonym was used in historical documents two hundred years ago. Use of exonyms may avoid the disorientation that might be caused by the juxtaposition of familiar exonyms in quotations with unfamiliar ethnonyms (names by which communities identify themselves in their own language) that refer to the same people. Ethnonyms also change over time. We cannot know for certain that all of the indigenous groups during the period under study used the same ethnonym as their descendants do today. Finally, not all groups are united on the question of how outsiders should refer to them.

I decided upon the term to use for each group after careful consideration, although the usage is not entirely satisfactory. The exonyms are generally corruptions of terms used by neighbors. For instance, Siouan and Algonkian bands referred to the Absarokes by using terms that traders translated as "Crows" or "Gens du Corbeau." In a pattern familiar throughout North America, Euroamericans adopted these terms rather than the ethnonym. Certainly some readers will consider my decision to adhere to these terms misguided. I hope they will not interpret them as disrespectful. In the interest of clarity I have listed some of the names by which ethnic groups are identified in this work, including other names by which they may be known and names that appear in historical documents and in scholarly literature. In some instances this required an interpretation that other scholars would dispute, but those cases are explained in the text. In accordance with the University of Oklahoma Press style, I use s for the plural of Indian groups (Crees, Crows, etc.).

♦ ♦ ♦

Arapahos: Inunainas (ethnonym); Gens des Vaches and its translation: Buffalo Indians; Tatood Indians; variations on Kanenavich and Caveninavish.

Assiniboines: Nakodas (ethnonym); variations on Assiniboines, Assinae Poets, Usinnepwats, Assinipualaks, and Sinepoets; Poets; variations on Stoney or Stone Indians.

Blackfoots (used to refer to the Piegans, Bloods, and Siksikas collectively): Nitsitapis (ethnonym); Blackfoot Confederacy; Gens du Large and its translation: Plain Indians; Archithinues, Slave Indians; Blackfeet; Peeagan tribes.

Bloods: variations on Kainais and Kainahs (ethnonym); Bloods or Bloody; Kennekoons.

Crees: Nahiawaks (ethnonym); variations of Crees, Krees, Kinistinaux, and Kilistinos; variations of Nehathawas, Nahathaways, and Ne-heth-aw-as; variations of Southern, Southerd, and Southd Indians.

Crows: variations on Absarokes, Absarokas, Apsarechas, Apsarukes, and Ererokas (ethnonym); Crows, Crow Mountain, and its translation Gens du Corbeau; Rocky Mountain Indians.

Flatheads: variations on Flatheads, Flatt heads, and Tetes Plattes; variations on Saleesh, Celish Indians; Tushapahs.

Gros Ventres: A'anis (ethnonym); variations on the English translation of Gros Ventres: Big Bellies, Paunch Indians; Gros Ventres of the Prairie; variations on Minatarees of the Prairies; Fall, Waterfall, or Rapid(s) Indians.

Hidatsas: variations on Big Bellies, Paunch Indians, and Gros Ventres; variations on Gros Ventres of the Missouri; variations on Mintaris, Minatarees, and Minetarees; variations on Wahtees, Ouachipouennes, and Naywatame Poets; Vault Indians; Willow Indians; Fall or Flying Fall Indians.

Kutenais: Tunaxas (ethnonym); variations on Coutonées, Cuttenchas, Coutenais, Cottonna haws, Kootenays, and Kootenais.

Piegans: variations on Peeagans, Pikunis, Peigans, Pekanows, Picanaus, Pee.ken.nows, Peeaganakoons, and Pikanis (ethnonym); Muddy River Indians or Muddies and occasionally Missouri River Indians; Blackfoots.

Sarcees: Sotinas, T'suu Tinas (ethnonym); variations on Sarcees, Sarsis, Circees, Sussus, and Sussews.

Shoshonis: variations on Shoshones, Sho Shones, and Cho shones; Snakes; variations on Alitans; variations on Kanasick thinewocks.

Siksikas: Siksikas (ethnonym); Blackfoots; Saxeekoons.

COMMON AND CONTESTED GROUND

INTRODUCTION

Beyond Culturalism

Communities in Contact

For most Whites throughout the past five centuries, the Indian of imagination and ideology has been as real, perhaps more real, than the Native American of actual existence and contact. . . . Although modern artists and writers assume their own imagery to be more in line with "reality" than that of their predecessors, they employ the imagery for much the same reasons and often with the same results as those persons of the past they so often scorn as uninformed, fanciful, or hypocritical.

—ROBERT F. BERKHOFER, JR.
The White Man's Indian

The history of the northwestern plains (see figure 0.1) between A.D. 200 and 1806 is not the story of cultural contact, cultural clash, cultural change, or cultural continuity. It would be wrong to reduce the history of the region and its people to these themes, which really represent current imagery. The reality was more complex, nuanced, and often paradoxical. For many centuries small human communities moved onto the northwestern plains and defended themselves there through complex combinations of trade, warfare, and diplomacy with neighbors who were often very unlike themselves. The northwestern plains were the common and contested ground of diverse communities.

The common and contested ground was a geographical region and a diplomatic and military reality. Despite the capricious and unforgiving

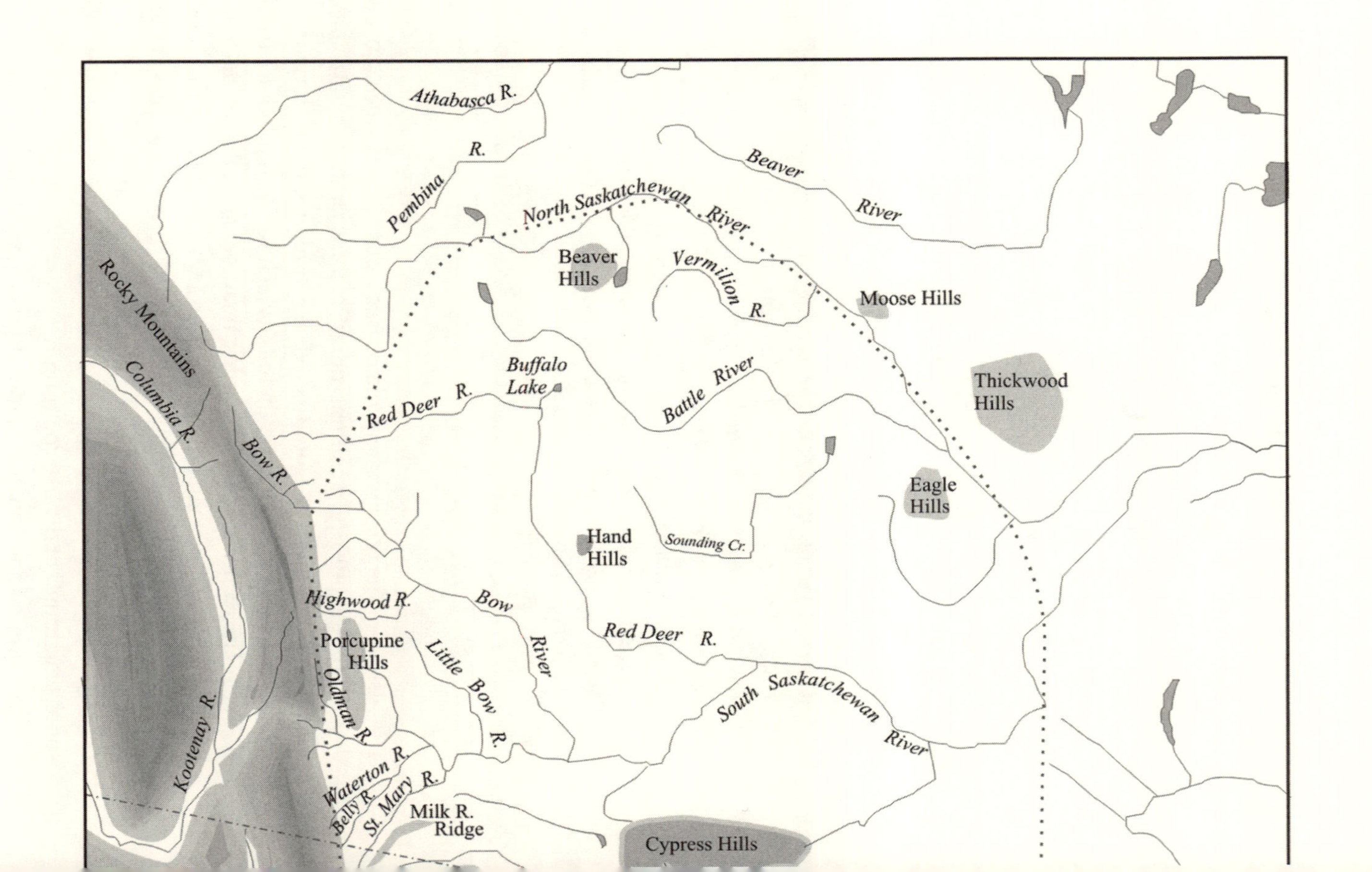

Athabasca R.
Pembina R.
Beaver River
North Saskatchewan River
Beaver Hills
Vermilion R.
Moose Hills
Rocky Mountains
Columbia R.
Buffalo Lake
Red Deer R.
Battle River
Thickwood Hills
Bow R.
Eagle Hills
Hand Hills
Sounding Cr.
Highwood R.
Bow River
Red Deer R.
Porcupine Hills
Little Bow R.
Oldman R.
Kootenay R.
South Saskatchewan River
Waterton R.
Belly R.
St. Mary R.
Milk R. Ridge
Cypress Hills

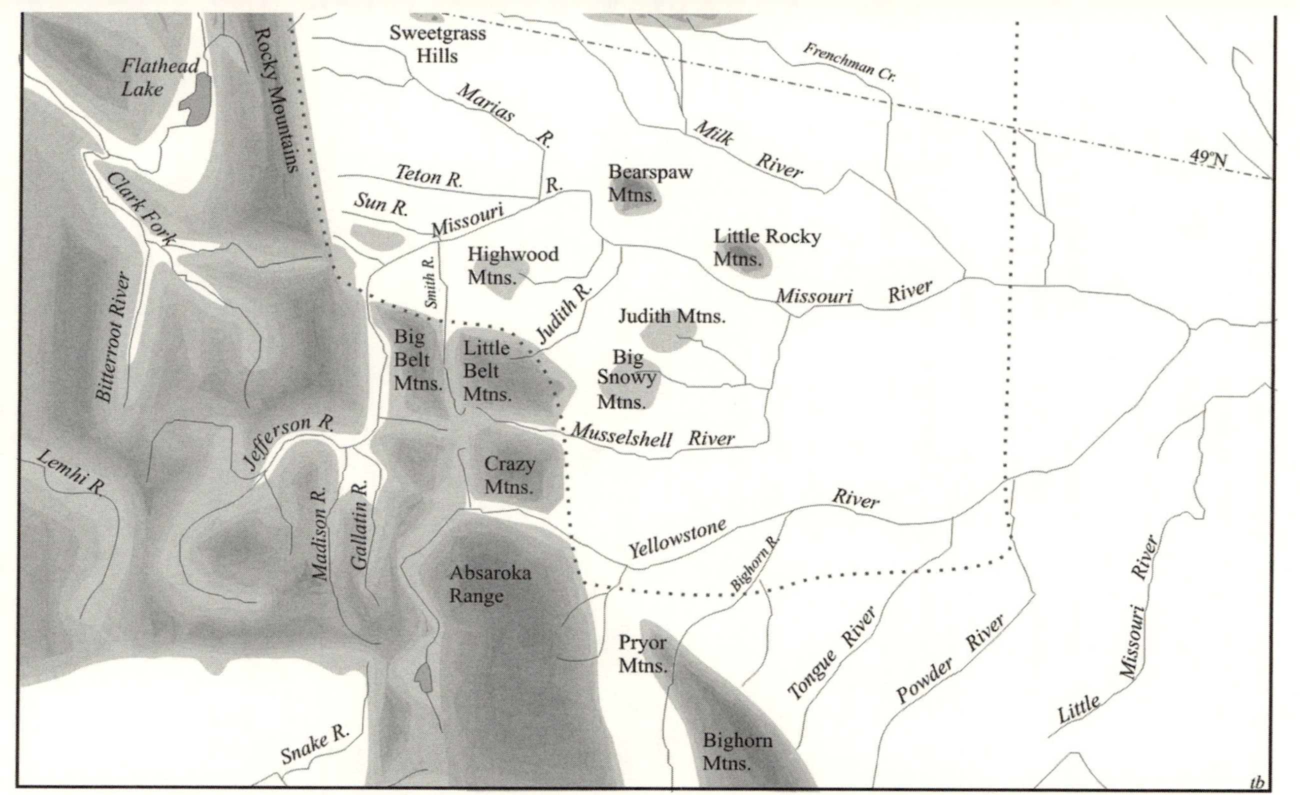

Fig. 0.1. The Northwestern Plains, Showing Boundaries of the Study Area. The 49th parallel represents the current boundary between Canada and the United States. (All figures by the author unless otherwise indicated)

climate, the northwestern plains' resources drew people from adjacent, less abundant areas. Because the region was so attractive to hunting societies, every community had to be willing and able to defend itself against rivals and to join with neighbors with whom it shared interests. Hunting bands continually cooperated and competed. Some moved onto the northwestern plains with little or no resistance from prior residents; others did so aggressively. Vulnerable groups coalesced with others to survive or fled.

When British, French, and Canadian newcomers arrived in the eighteenth century, they conformed to these patterns even though their arrival dramatically changed life in the region. Euroamerican traders lived in separate but shared worlds with the aboriginal inhabitants. The traders struggled, often violently, among themselves, yet they came as both arms brokers and self-styled peacemakers to the North American Indians. Despite their own antagonisms, Euroamericans often built competing trading posts within single stockades in order to defend themselves against the Indians. Then they made common cause, and often intermarried, with Indians. The newcomers held no coercive power over their trading partners, yet they unleashed powerful forces that influenced every inhabitant of the region. By 1806 they felt less confident in the crucible of the northwestern plains than they had fifteen years earlier. A middle ground, even a middle people, developed between Indians and Euroamericans; more significantly, Euroamerican traders became important but militarily weak participants in the complex patterns of trade, warfare, politics, and diplomacy.

This book departs from the themes that have dominated the scholarly study of the indigenous history of the region for a century. From the 1880s to the 1950s anthropologists like Horatio Hale, Alexander Chamberlain, Clark Wissler, Alfred Kroeber, and Robert Lowie who studied the American Indian societies of the northwestern plains were on the leading edge of their discipline. At a time when anthropologists were developing and popularizing the concept of culture, indigenous cultures were their preoccupation. These "salvage anthropologists" assumed that they were racing against time, as the rush of civilization threatened to eradicate the vestiges of what they believed were pristine pre-Columbian cultures. They sought to preserve, record, and describe aspects of indigenous cultures before they were lost or forgotten. Meanwhile, historians like Harold Innis, A. S.

Morton, Hiram Chittenden, and Walter Prescott Webb wrote their classic works. In contrast to the anthropologists, these scholars focused on the Euroamerican dimensions of the past. Even in their histories of the so-called fur trade, they gave American Indians only cursory treatment.

Oscar Lewis's *The Effects of White Contact upon Blackfoot Culture* (1942) marked an important departure in the anthropological scholarship of the northwestern plains. As the title indicates, his work, like that of his predecessors, centered on questions of culture. Lewis, though, was very critical of what he considered the failure of anthropologists to understand cultural change historically. Foreshadowing the formal emergence of the field of ethnohistory, he argued that "the intensive search for such materials [historical documents] and their exhaustive and critical analysis can help bridge the gap between the disciplines of history and anthropology." Relying overwhelmingly on published primary sources, Lewis contended that "the fur trade was the mainspring of Blackfoot cultural change.... [I]t was the fur trade together with the horse and the gun which had a dynamic effect upon Blackfoot institutions."[1] Because it uses historical documents to address anthropological questions, Lewis's study ranks as one of the earliest works in ethnohistory.

Soon afterward another plains anthropologist, Frank Raymond Secoy, used historical documents to answer historical questions. His book *Changing Military Patterns of the Great Plains* (1953) emphasizes the military rather than the cultural repercussions of the arrival of the horse and the gun. While Lewis's work pioneered in ethnohistory, Secoy's study remains unique. The only comparable work for the northwestern plains is John Milloy's *The Plains Cree* (1988). Anthony McGinnis's *Counting Coup and Cutting Horses* (1990) does not displace Secoy's book as the most valuable general study of the history of warfare on the plains.

Stimulated by the emergence of the journal *Ethnohistory* in 1954, ethnohistory emerged as an identifiable field of study in the 1950s. Ethnohistorians try to use historical and anthropological sources, methods, and approaches to reconstruct aspects of past cultures and to trace cultural change and continuity. In the 1950s and early 1960s John C. Ewers kept the scholarship on the history of the northwestern plains at the leading edge of the field of ethnohistory. Based on interviews with Piegan and Blood informants born in the 1850s, 1860s, and 1870s and extensive research in primary sources, including unpublished documents, his studies were the very model of ethnohistorical writing. Ewers's magnum opus, *The Blackfeet: Raiders on the Northwestern Plains* (1958), remains the standard history of the Blackfoot bands from their "Dog Days" to the early reservation era.

As ethnohistory flourished in the 1950s, ethnohistorians and "New Indian historians" turned away from the history of the northwestern plains and toward topics relevant to North American Indian land claims and regions where they believed the documentary record was particularly strong. Several important, though not groundbreaking, studies relevant to the northwestern plains have been published since 1958. John Milloy's *The Plains Cree* (1988; actually completed in 1972) focuses on trade, diplomacy, and warfare, although it depicts ethnic groups as far more cohesive than the evidence allows. Loretta Fowler's fine study *Shared Symbols, Contested Meanings* (1987) includes a brief survey of Gros Ventre history before 1806, but it emphasizes cultural themes and the reservation era. Dale Russell's *Eighteenth-Century Western Cree and Their Neighbours* (1991) presents a great deal of evidence on the Cree, Assiniboine, Gros Ventre, and Blackfoot bands in the eighteenth century, but it does not address the issues examined in this book. Frederick Hoxie's *Parading through History* (1995) interprets Crow history after 1806.

After 1970 scholars turned toward Euroamerican-Indian trading relationships. Arthur J. Ray set the standard. His *Indians in the Fur Trade* (1974) explores the history of Indian-newcomer trade in the prairie-forest ecotone northeast of the Great Plains and deals only peripherally with the northwestern plains. Everyone interested in the history of American Indian roles in post-Columbian trading systems should be familiar with this book. Environmentally defined studies are rare today, despite their obvious potential. Ray showed how significant the ecological diversity and natural cycles of the prairie-forest ecotone were in its human history. Strangely, although many acknowledge the importance of Ray's work, few seem to have recognized the value of its organizing scheme or to have emulated it. David Wishart's *The Fur Trade of the American West* (1979) deals more directly with the northwestern plains, but it focuses on Euroamericans in the trade, beginning in 1806.

At present the scholarly literature on the history of the northwestern plains, although well developed, has important lacunae. Groundbreaking anthropological and ethnohistorical studies published before 1960 contributed tremendously to our knowledge of northwestern plains history and influenced scholarship elsewhere. But they are now dated. Anthropologists like Ewers, Secoy, and Lewis did not consult the vast resources of the Hudson's Bay Company Archives (HBCA). Furthermore, historical and anthropological literature published since 1960 undermines some of the assumptions and conclusions of these older studies. Historical works published before 1970, which dealt largely with the history of newcomers in the region, are also dated. Newer studies have partially updated the literature, but many of these works reflect the same scholarly themes that have dominated the literature since the 1880s. The strengths and gaps in the literature as a whole remain the same as they were forty years ago.

Much of the literature on the northwestern plains addresses the history of ethnic groups, deals with cultural themes, and emphasizes the cultural determinants of human behavior. We are endowed with fine histories of several ethnic groups, most of which show how various cultures exhibited continuity and dynamism over time. The literature also reflects a concern with the history of Indian-newcomer relations. Scholars often debate how contact with the Western world affected indigenous cultures. Their answers have changed dramatically over the years, but the central questions are largely the same. It is useful to address some of the evidence, themes, and questions that the scholarship often neglects.

As important as the arrival of Westerners and Western goods and ideas was, we should not assume that Indian-newcomer relations were at the center of Indian life on the northwestern plains before 1806. Euroamericans forged very important bonds with some of their Indian trading partners, especially Cree and Assiniboine bands. Relations with others, like the Blackfoot and Gros Ventre bands, were generally restricted to brief trading encounters. To the Shoshonis, Crows, Flatheads, and others, Westerners were still strangers in 1806. We cannot hope to understand Indian-Euroamerican relations adequately unless we attempt to explain them in the context of broader Indian interactions. Any middle ground between indigenous peoples and newcomers developed within the common and contested ground. Euroamericans became influential but not powerful participants in the ancient, dynamic, and complex patterns of trade, diplomacy, and warfare in the region. This book focuses on the history of human interaction generally, rather than on Indian-newcomer relations specifically, emphasizing the political, diplomatic, military, and environmental dimensions of this history.

It is also important to reconcile the significance of event and individual with the importance of structure and culture. Ethnohistorians usually seek to describe and explain gradual change in fundamental aspects of culture over the long term. They rarely discuss prominent individuals or dramatic turning points. This is not only because of the weaknesses in the documentary evidence, but also because anthropological history tends to downplay the role of individuals and events in history—as if events and individuals were surface disturbances or froth on the great tides of history. Anthropologist Marshall Sahlins, one of the most prominent critics of structuralist scholarship, has struggled against the false dualism between structure and circumstance. An event, according to Sahlins, is "at once a sui generis phenomenon with its own force, shape, and causes, and the significance these qualities acquire in the cultural context."[2] A historical approach requires attentiveness to the role of events and individuals in the past.

Certain individuals greatly influenced the course of events on the northwestern plains. Admittedly the documentary evidence is quiet for the period before 1806. For instance, most of the traders' documents were addressed to Hudson's Bay Company (HBC) officials in London who did not want and could not have benefited from detailed information about individual Indians. Still, correspondence among fellow traders frequently does mention individuals and shows that traders made it their business to get to know Indians individually. During the winter of 1778–79 British surveyor Philip Turnor observed the HBC traders at work in the Lower Saskatchewan River region. He noted that "there is not an Indian in that part of the Country but Mr Willm Tomison or Robt Longmore [Longmoor] or both of them is acquainted with and many of them not known by any other of Your Honors Servants and the Indians seem very fond of them both."[3] Traders cultivated the friendship and respect of Indian leaders, for success depended on their cooperation. Despite gaps in the documentary evidence, scholars must recognize that careful attention to individuals and events will clarify our understanding of the past.

Most importantly, this book is intended to dispel the mistaken impressions sometimes engendered by tribal histories, which tend to suggest that cultural units or ethnic groups corresponded to social, political, and economic units. Criticism of the concepts of "tribe" and "culture" began in earnest during the late 1960s.[4] In 1974 Susan R. Sharrock argued that "the concept of 'tribe' has acted as a Procrustean bed for the generation of interethnic theory, and that theory is inadequate to explain the relationships among ethnic units that are historically documented."[5] Recently some scholars have indicated a resolve to move beyond the traditional emphasis on cultural groups. That is one of the great contributions of Richard White's *The Middle Ground*.[6] White and others have shown the limitations of tribal histories and the value of studies that recognize the important relationships within and among ethnic groups.

We can understand relations among indigenous communities only if we appreciate that North American Indian bands were complex organized societies. Band societies were not "primitive" or "simple" societies, akin to humanity "in the state of nature."[7] Band societies may appear deceptively simple, even unorganized, because they lack the formal institutions that state societies possess. But band societies achieve through informal means exactly what state societies accomplish in other ways. Indeed the flexibility, fluidity, and informality of band societies enabled them to respond quickly and effectively to the rapidly changing circumstances that they typically faced.

Bands on the northwestern plains, as elsewhere, were organized around extended families. In many ways these families functioned as extended families do anywhere. People rose to prominence and held these positions based on their reputations among other members of a band and, to a degree, among members of affiliated bands. Leaders carried no formal title and no power, only influence in proportion to their reputations. The following passage from the journals of Captain Meriwether Lewis and William Clark on leadership among the Shoshonis also applies to band societies generally:

> the authority of the Chief [is] nothing more than mere admonition supported by the influence which the propriety of his own examplery conduct may have acquired him in the minds of the individuals who compose the band. the title of chief is not hereditary, nor can I learn that there is any cerimony of instalment, or other epoch in the life of a Chief from which his title as such can be dated. in fact every man is a chief, but all have not an equal influence on the

> minds of the other members of the community, and he who happens to enjoy the greatest share of confidence is the principal Chief.[8]

While band leaders wielded no coercive power, they greatly influenced followers in their own bands and often in other bands, even across ethnic boundaries.

A band member was remarkably free to leave one band and join another. This fluidity did not threaten but enhanced the communities' stability. Poorly equipped to deal with conflict and division, bands used many informal means to arrive at consensus. Dissenters, for instance, were encouraged to acquiesce rather than to agitate when they disagreed with the majority of band members. If they did not accept the decision of the majority, they could vote with their feet by joining another band either temporarily or permanently. They could easily do so when several bands separated after camping together for a time. Every band member inevitably had family members in other bands, so the move from one band to another often would not be difficult. Unhampered movement meant that local bands could emerge, grow, wane, and disappear over time. Despite apparent individualism, bands also exhibited a strong sense of community. Individuals knew that they could not survive long unless they were members of a supporting community. No community wanted troublemakers, so wise individuals learned to place the needs of the community ahead of personal interests. Band societies reserved some of their harshest sanctions for selfish behavior. Only those with established reputations of generosity could ever expect to become prominent.

Just as individualism and communitarianism existed in tension, so did band autonomy and broader allegiances. Especially during the equestrian era bands needed to cooperate with others to achieve common aims. Occasions for combined action ranged from communal bison hunts and religious ceremonies to warfare. No band could survive long on the northwestern plains without the friendship and assistance of neighbors. Local bands were tied to other bands in several ways. Ethnicity was an important unifying force. For instance, the fact that they spoke the same language, shared similar beliefs and customs, and shared a common history tied all Crow bands together. Their destiny was also linked because the behavior of a particular band affected the others. After an attack by one Crow band a Blackfoot band might retaliate against any of the other Crow bands. This meant that the relationships between all Crow bands and all Blackfoot bands tended to be similar. Certain Crow leaders and bands could break established patterns, but any single band would have found it very difficult to pursue an autonomous policy toward the Blackfoot bands over the long term. The same principle influenced relations among all ethnic groups.

All bands were also linked to others by kinship networks. Bands of the same ethnicity were naturally tied together in this way. So members of any Shoshoni band always had brothers, sisters, parents, children, aunts, uncles, and relations by marriage in other Shoshoni bands. Kinship entailed reciprocal obligations. When one band was unsuccessful in the hunt or feared attack by enemies, it could count on the assistance of neighboring kin. Family helped family.

Kinship ties also crossed ethnic lines, however. By focusing on a single group such as the Crees, the Kutenais, or the Crows, we risk overlooking the important network of relationships that existed between ethnic groups. Scholars who define their studies culturally almost inevitably select culturally specific issues. Intelligent readers can be forgiven for believing that the boundaries between cultural groups were relatively solid and impermeable. They could easily conclude that a Cree woman almost inevitably lived in her homogeneous Cree band until she married a Cree man of another band. Their children would certainly be raised as Crees. These assumptions would be mistaken.

The documentary record for the northwestern plains offers ample evidence of contact, mixing, merging, and amalgamation among cultural groups. It was routine for a single encampment to include members of several local bands belonging to several cultures. So when Anthony Henday met a large encampment of Indians in 1754, it probably included Blood, Blackfoot, and Piegan as well as Gros Ventre bands. During the winter of 1792–93 Sakatow's Piegan band camped together with Blood, Siksika, Sarcee, and Cree bands at the same time. Soon afterward his band also visited with some Shoshonis and Kutenais. For the most part the scholarly literature does not reflect the fact that combined encampments (Crow-Shoshoni, Shoshoni-Flathead, Cree-Sarcee, Sarcee-Blood, Siksika-Assiniboine, and Assiniboine-Cree) were normal. This is odd, for traders' journals often mention mixed bands. Daniel Williams Harmon wrote of the northern plains that "all neighboring tribes frequently intermarry."[9]

The northern plains were not unique. John Moore has explained: "I do not know of any kinship schedules, or explicitly stated patterns of exogamy, that are predicated on marriage with foreigners, although it takes place in all societies I know about."[10] Evidence from around the world has shown that "many band societies are bi- or multilingual and highly diverse in regard to general cultural content."[11] Where individuals and bands of different ethnic groups frequently mixed, intermarried, and circulated, boundaries between them could become indistinct. Long-term friendships among Algonquian-speaking Cree and Siouan-speaking Assiniboine bands of the northern plains produced a merged identity. Some bands were neither Cree nor Assiniboine, but interethnic.[12]

Interethnicity was not incidental to life on the northwestern plains; it was essential. Distinct and relatively autonomous communities often camped together temporarily, allowing them to make or renew friendships, trade, hunt, or go to war together. Interethnic contact was not limited to these temporary encounters. Boundaries between indigenous groups were always permeable and occasionally indistinct. Little prevented an individual born within one ethnic group from becoming a prominent leader in another. The frequency with which this happened suggests that interethnicity was very common and that persons with a mixed heritage were well suited to leadership roles, especially in contexts where diplomacy became important.[13]

Recognition of the complexity and fluidity of band societies opens the way to new insights, but it also presents some difficulties. We know that people belonging to a single indigenous culture often disagreed sharply because they held different positions in society, because they had different personal priorities, or because they had different perceptions of situations that they were confronting. We also know that people of diverse cultures could work together either intermittently or continuously because each group believed that doing so would serve its best interests. Unfortunately, only rarely has enough evidence survived for us to reconstruct the history of nonliterate peoples at these levels. But by acknowledging that the past was far more complex than the historical documents suggest we become more alert when we encounter evidence that does shed light on particular individuals, milestone events, interest groups, intraethnic dynamics, and interethnic relationships. Then we can begin to explore how a myriad of forces shaped the history of a region.[14]

Band societies differed markedly from Western societies in many ways that few Euroamerican traders ever understood fully. Relations between human communities on the northwestern plains exhibited much of—and owed much to—the complexity, fluidity, and flexibility that were unique and essential to the North American Indian bands. For this reason, we are wise to describe patterns of Indian interaction with terms that do not carry misleading connotations. The term "alliance," with its connotations of established protocol, permanence, and formality, inaccurately describes relationships on the northwestern plains, regardless of how friendly or long-standing they were.

Terms such as "coalition," "affiliation," and "association" better capture the essence of these relationships. Coalitions, even if durable, were fluid and temporary, in need of constant renewal and maintenance. They were expedient combinations in which distinct and autonomous groups worked toward specific aims, but which did not necessarily entail reciprocity. Even within long-term coalitions bands could and did endure hostile incidents, even bloodshed and death. If one partner in a coalition was much weaker militarily than another, it would be vulnerable to occasional raids by its partner. Conversely, long-standing enmities could be punctuated by peaceful and cooperative encounters. "Inveterate foes" sometimes not only engaged in peaceful trade but joined to wage war on a third party that might normally be at peace with one of the others. This might imply no efforts at long-term rapprochement; two normally hostile bands might simply have judged that at a certain time peaceful relations, even cooperation in warfare, suited their interests better than warring against one another did.

The history of the northwestern plains is not primarily a cultural story. Communities of different traditions, aboriginal and newcomer, often coalesced when their interests coincided, while communities with common lifeways were often riven by factionalism and competing interests. It is worth remembering that anthropological definitions of culture developed only in the twentieth century and that in recent years many scholars have become critical of culture concepts. Some have recommended dispensing with the idea of culture altogether. Although anthropologists and ethnohistorians employ culture concepts in sophisticated ways, there is no reason to believe that culture, however it is defined, was the primary characteristic by which people on the northwestern plains identified themselves before 1806. Lines of friendship and animosity, kinship and influence, were not defined merely by culture. Neither is there evidence that the indigenous inhabitants of the northwestern plains were concerned about cultural change or continuity in that period. Communities did come into contact; they cooperated and clashed with each other, and their lifeways exhibited both change and continuity. But cultural contact, clash, change, and continuity are the dominant themes of twentieth-century literature in indigenous North American history; they were not the preoccupations of the historical actors themselves.

CHAPTER ONE

"A Good Country"

The Crow country is a good country. The Great Spirit has put it exactly in the right place; while you are in it you fare well; whenever you go out of it, whichever way you travel you fare worse. If you go to the south, you have to wander over great barren plains; the water is warm and bad and you meet with fever and ague. To the north it is cold; the winters are long and bitter and there is no grass; you can not keep horses there but must travel with dogs. What is a country without horses? On the Columbia they are poor and dirty, paddle about in canoes and eat fish. . . . fish is poor food. To the east they dwell in villages; they live well, but they drink the muddy waters of the Missouri—that is bad.

—ATTRIBUTED TO ARAPOOASH (SORE BELLY)
ca. 1830

Farmers who watched their crops and dreams wither in the sun and wind during the 1930s would have scorned Sore Belly's appraisal of the northwestern plains. Many of the homes they abandoned decades ago now stand as silent bleached monuments to the hopes that died there. Even today the core of the region remains a vast, nearly treeless and sparsely populated area of large ranches and small towns. Substantial cities have grown only along its more forgiving margins. Those who stayed and prospered never seem to have lost their awe of the stark landscape, the immense but miserly sky, and the fierce climate of violent contrasts.

It is a landscape dominated by apparently lifeless shades of yellow, brown, and white. Greenery passes quickly. In both summer and winter the sun and the wind are relentless. The January sun is a mere irony; the July sun defies all life. And the nearly ceaseless wind makes the climate of the northwestern plains what it is, subjecting the region to the most sudden weather changes on the globe. Chinook winds quickly drive midwinter temperatures well above freezing. Equally impressive, and far more dangerous, are the northerly winds that suddenly plunge the region into a protracted deep freeze.

Today irrigation supports intensive agriculture even on some of the driest portions of the northwestern plains. In other areas dryland farmers coax harvests from arid soils, but sprawling cattle ranches now dominate the core of the plains. Why did Sore Belly speak so highly of the land that so many Euroamerican migrants fled? Euroamericans brought traditional and ill-suited eastern agricultural practices and technologies to the high plains, while Sore Belly's ancestors, also migrants from agricultural societies in the east, never made a significant effort to import agricultural methods to the region. Sore Belly's people valued the northwestern plains for their bison. Documentary evidence and scientific literature suggest that because of their topography, vegetation, and climate the northwestern plains supported exceptional numbers of bison and other large ungulates and simplified the task of nomadic human hunters who depended upon them. Hunting societies learned how to manage the environment to enhance its ability to support bison.

Douglas Bamforth's *Ecology and Human Organization on the Great Plains* (1988) is one of the most useful ecological studies of human history on the Great Plains. It demonstrates the importance of understanding a region's environment in order to appreciate that region's human history. Bamforth argued that during the bison era the bison enjoyed more abundant forage on the northeastern plains than in any other area of the plains and that pasturage generally decreased to the south and west, with the crucial exception of the Black Hills. Bison were more plentiful, the herds more evenly distributed, and their movements more predictable on the northeastern plains than anywhere else. These environmental factors allowed human hunters to form generally larger and more sedentary communities on the northeastern plains than they could elsewhere on the Great Plains. Since larger communities could usually dominate smaller ones militarily,

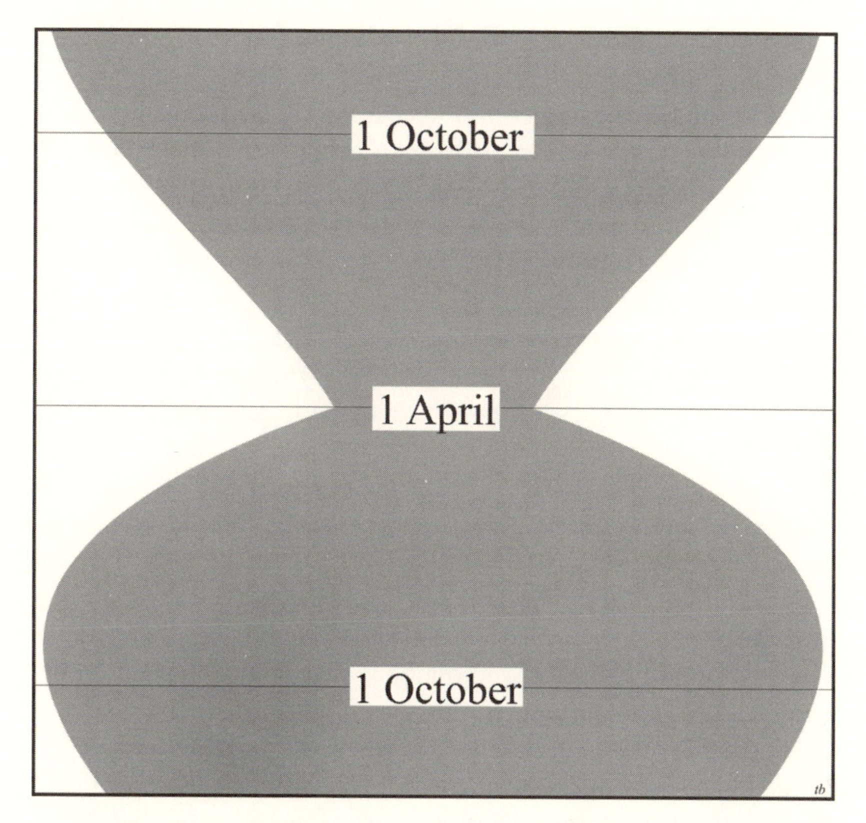

Fig. 1.1. The Bison Hourglass. The shaded area suggests how bison numbers changed during a typical year. Food supplies declined during the fall and winter, producing an hourglass through which only a certain portion of the bison population could fit. The bison hourglass had important implications for the human population in a region.

northeastern groups tended to supplant southwestern groups. The Dakotas were an exception that proved the rule: sustained by the bison oasis of the Black Hills, they dominated the west-central plains.[1]

Bamforth wisely emphasized that the bison density in an area was not limited by the region's total yearly forage production but by its carrying capacity during the leanest period of the year. The amount of food available to bison in any environment varied according to the seasons. A graph of these variations resembles an hourglass (see figure 1.1). Most of the grass growth on the world's grasslands occurs in a relatively short period. During that time herbivores have more food than they can possibly eat.

As grasses go dormant, however, they can support far fewer animals. On the northern plains the hourglass narrowed during much of the summer and all of the fall and winter, with the narrowest part of the hourglass coming just as the grasses sprouted in early spring. The width of the hourglass at its narrowest, not its widest, was crucial. Bison numbers varied seasonally in much the same way: populations grew rapidly during the peak calving period in spring but began decreasing in the summer and continued to drop until the spring, as predators, disease, natural hazards, and starvation culled the herds. The average size of a bison population in any area depended largely on the number of animals (particularly reproducing cows) that could squeeze through the neck of the hourglass in late winter.

In order to understand the neck of the hourglass, it is important to examine the total nutrients, not the total forage, available to bison at the leanest time of the year. The distinction is crucial. Forage abundance can differ significantly from forage quality, particularly in late winter and early spring, when the grasses have not grown for about seven months. The northeastern plains produced more forage than other areas of the great plains, but for various reasons forage on the northwestern plains was of considerably higher quality during the critical period of the year.

The topography of the northwestern plains (see figure 0.1) offered certain advantages to bison and the humans who hunted them. The terrain is generally undulating, but it is roughest and highest near the Rocky Mountains. A low divide runs approximately east-west near the present-day United States–Canadian border. Rivers to the north of this divide flow into the Saskatchewan River system and ultimately into the Hudson Bay; those to the south flow into the Missouri River and then into the Gulf of Mexico. Although the climate of much of the region is semi-arid, its substantial rivers are fed by the snows of the Rocky Mountains, and the uneven terrain forms catchments for runoff. The large rivers and even some of the minor ones occupy wide and deep sheltering valleys originally sculpted by the meltwater channels at the end of the last ice age. These broad valleys support luxuriant grasses and stands of cottonwood (*Populus deltoides*) and deciduous shrubs.

Remarkable highlands also dot the landscape, including the Hand Hills, Cypress Hills, Sweetgrass Hills, Bearspaw Mountains, Highwood Mountains, Big Belt Mountains, Little Belt Mountains, Judith Mountains, and Snowy Mountains. They rise from a few hundred to a few thousand feet above the surrounding plain. Most are high enough to cause orographic precipitation, particularly in May and June, when cool, moist northeasterly winds are common. For example, in most years the Cypress Hills get two or three more inches of precipitation during the growing season than do the surrounding plains.[2] The extra rainfall and cooler summer temperatures (resulting in less evaporation) enable the highlands to support robust grass growth.[3] Some of the higher hills, like the Cypress Hills, Judith Mountains, and Bearspaw Mountains, even supported substantial forests of lodgepole pine, jack pine, white spruce, and douglas fir. Spring heads and creeks in the highlands and in the foothills of the Rockies provided water for foragers when precipitation was scanty.[4] The rough topography of the northwestern plains, with its sheltered areas, catchments, and springs, provided excellent conditions for large herds of bison for the same reason that it provides excellent range pasture for cattle today.

The unique vegetation of the northwestern plains also helped support large bison herds during the winter. The northern mixed prairie, one of the main grassland regions of the Great Plains, dominated the northwestern plains (see figures 1.2 and 1.3). The northern mixed prairie (sometimes misleadingly lumped together with the shortgrass plains farther south, which are dominated by warm-season grasses) consists of both warm- and cool-season grasses. This is important because the cool-season grasses begin growing earlier in the spring than warm-season grasses do and often enter a second growth in the fall.

The northern mixed grassland region included two subregions, the xeric (dry) mixed grasslands and the more productive mesic (moist) mixed grasslands. In three directions the northern mixed prairie merged with the relatively luxuriant fescue prairie. The boundaries of these regions and subregions were influenced by topography but determined primarily by the ratio of precipitation to evaporation.[5] Each grassland region blended imperceptibly into the next as climatic and topographic conditions varied. Highlands or north-facing slopes in the dry mixed prairie region supported grasses more typical of the moist mixed prairie or the fescue grasslands, while south-facing slopes in the fescue prairie region resembled the mixed prairie. The border between the moist and dry prairies was all the more imperceptible because all the grasses grew

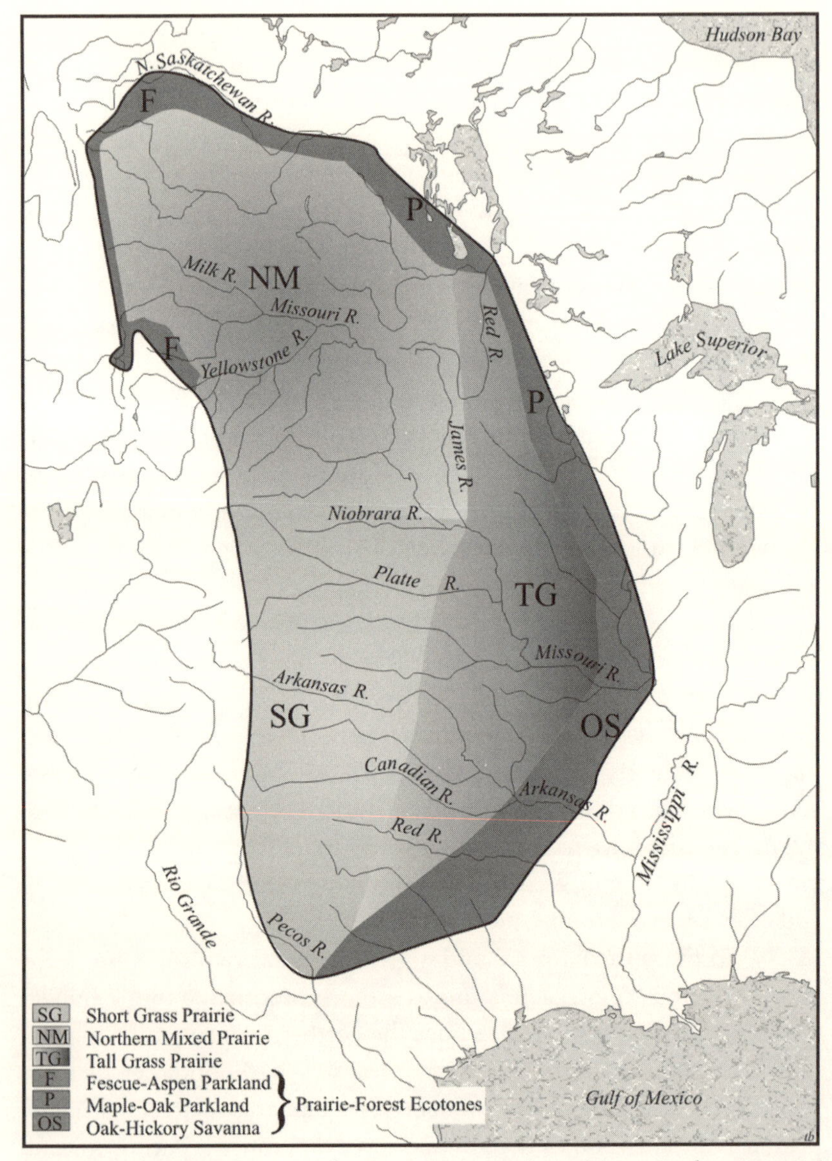

Fig. 1.2. Grassland Regions of the Great Plains. Adapted from Kücher, *Potential Natural Vegetation*.

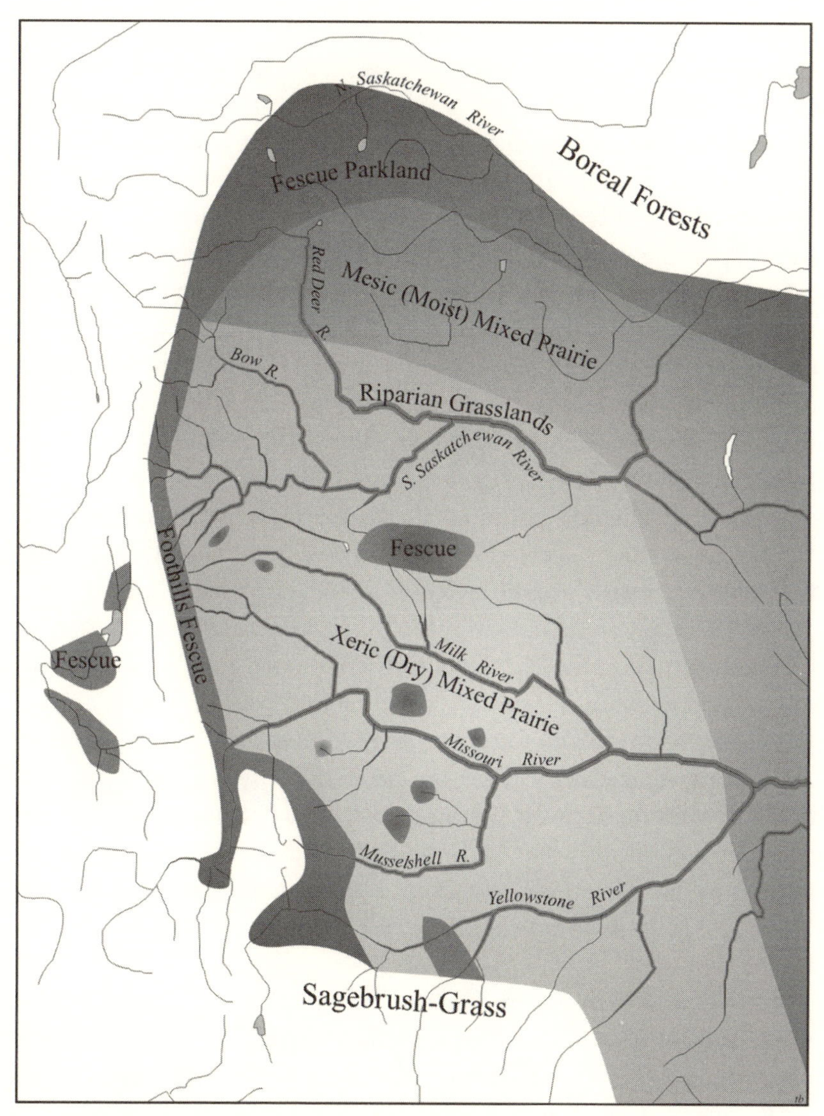

Fig. 1.3. Grasslands of the Northwestern Plains. Adapted from Kücher, *Potential Natural Vegetation*; Payne, *Vegetative Rangeland Types*; and Strong and Leggat, *Ecoregions of Alberta*.

in both areas, but most grew taller where moister and cooler conditions allowed. None of these grasslands was as productive as the tallgrass prairies to the east, but the neck of the hourglass may have been wider on the northwestern prairies than anywhere else on the Great Plains.

In the dry subregion the growing season began as early as late March or early April south of the Cypress Hills (not including the hills themselves), peaking in May and June, and then waned as the soils dried and the temperatures rose in summer. The effects of climate and elevation produced an earlier start to the growing season than is found farther north, east, or west. This meant that the critical period of late winter and early spring, when bison were under the greatest stress, ended earlier here than it did elsewhere on the northern plains. Growth was sparse, but the tender new shoots were highly nutritious.

While a considerable portion of the grasses in the dry mixed prairie in this region are cool-season grasses like needle-and-thread (*Stipa comata*) that end their year's growth by early July, the dominant species is the drought resistant, warm-season grass, blue grama (*Bouteloua gracilis*). Blue grama and buffalograss (*Buchloë dactyloides*) sprout a month after the other grasses and in some areas may grow until mid-August. The dry mixed prairies are relatively unproductive, and warm-season grasses are generally less nutritious than cool-season species. Still, the cool-season grasses of the dry northern mixed prairie sprouted exceptionally early, and its warm-season grasses recovered well from the heavy grazing they experienced in late spring and summer.[6] Blue grama grass, the dominant warm-season grass in the region, is exceptionally nutritious and was a favorite of bison.[7]

The moist subregion of the northern mixed prairie lay generally to the north of the dry subregion. The dominant species are various heavier-yielding cool-season mid-grasses, including western porcupine grass (*Stipa curtiseta*), western wheatgrass (*Agropyron smithii*), needle-and-thread, and the very early June grass (*Koeleria cristata*). Although the cool-season grasses in the moist prairie sprouted at least a week later in the spring than they did on the dry prairie, in the wetter, cooler conditions they remained green longer and later. On the whole the moist grassland was substantially more productive. Because early-growing cool-season grasses dominated the moist prairie, it usually had twice the carrying capacity of the dry prairie by the end of April.[8]

The fescue (*Festuca* spp.) grasses were the most productive species on the northwestern plains. The region provided fine bison grounds, because herds were most reliant upon fescue grasses during the winter. These grasslands, which were dominated by foothills rough fescue (*F. campestris*) or its smaller cousin, plains rough fescue (*F. hallii*), throve on the cooler temperatures (resulting in lower evaporation) and wetter conditions that prevailed along the margins of the prairies. A crescent of fescue grasslands formed a nearly unbroken arc in association with aspen (*Populus tremuloides*) groves roughly from the forks of the Saskatchewan River west along the northern rim of the plains (the parkland), continuing as a narrow band of submontane vegetation along the foothills of the Rocky Mountains along the western margins of the plains and then into the Big Belt, Little Belt, and Snowy Mountains of present-day Montana. Foothills rough fescue also penetrated the mountains in major valleys such as the Bow River, Crowsnest River, Waterton River, and Missouri River valleys, where it occasionally occurred near subalpine meadows.[9] Fescue grasslands were also scattered throughout the northwestern plains on highland areas, on benchlands, and occasionally on north-facing slopes in the mixed-grass prairie. The Milk River Ridge, Sweetgrass Hills, Cypress Hills, and Bearspaw Mountains supported rough fescue in almost pure stands that rivaled the northern parkland in their productivity.[10]

Duncan McNab McEachran, a visitor from eastern Canada in 1881, described the quality of pasturage and shelter in the fescue grasslands of the foothills in the upper Bow River valley at the northern end of the Porcupine Hills:

> There is an abundance of pine and cottonwood on Jumping Pound Creek and the hillsides, besides numerous thickets of alder and willow scattered here and there over the range, which afford excellent shelter for stock in winter. The grasses are most luxuriant, especially what is known as "bunch grass" [foothills rough fescue], and wild vetch or peavine [*Vicia americana* or *Lathyrus venosus*], and on the lower levels, in damper soil, the blue joint grass, which resembles the English rye grass, but grows stronger and higher [probably reed grasses, *Calamagrostis* spp.]. On some of the upland meadows wild Timothy [perhaps alpine timothy, *Phleum alpinum*] is also found. These grasses grow in many places from one to two feet high, and

> cover the ground like a thick mat. Nowhere else has the writer seen such abundance of feed for cattle.[11]

Regarding the Oldman River valley on the southern extremes of the Porcupine Hills, he wrote that

> here as further north, we found that as we neared the mountains, owing to the atmosphere being more moist and rains more frequent, the grasses became more luxuriant, and for about twenty-five miles from the base of the mountains, including the foothills, there is an inexhaustible growth of rich nutritious grass. It some places it is so thick and so long as to impede the progress of the horses.[12]

McEachran visited when bison no longer grazed these lands and before ranching grew to major importance, so the grasses were probably thicker than they were under bison grazing; but his descriptions hint at the productivity and nutritiousness of the foothills grasses. George M. Dawson of the Canadian Geological Survey echoed this opinion, noting that "within the Porcupines and their northward and southward extensions some of the best cattle ranching country of the entire North-west is situated."[13]

Early observers wrote as enthusiastically of the other highlands on the northwestern plains. Captain John Palliser, when visiting the Cypress Hills during the dry late 1850s, described them as "a perfect oasis in the desert" marked by an "abundance of water and pasture."[14] When John Macoun visited the Cypress Hills in the much wetter early 1880s, he noted that "no better summer pasture is to be found in all the wide North-west than exists on these hills, as the grass is always green, water of the best quality is always abundant, and shelter from the autumnal and winter storms always at hand."[15] Captain W. J. Twining, chief astronomer and surveyor for the United States during the boundary survey of the 1870s, wrote of the Sweetgrass Hills region that "these Three Buttes are the centre of the feeding ground of the great northern herd of buffaloes. . . . The number of animals is beyond all estimation."[16] Dawson noted that copious springs supported wooded valleys and grassy foothills in the hills.[17] Even smaller hills, like the Hand Hills, supported richer vegetation than the surrounding plains. According to Palliser, "the plain all around the base of the [Hand] hills is bare and arid, but the high level of the hill bears a fair and almost rich pasture, being 680 feet higher than the plain; it also contains lakes of pure fresh water, and gullies with a small growth of poplar."[18]

Like the foothills and uplands, the wide valley bottoms and depressions were important to bison during the winter; like the fescue grasslands, they supported heavy grass growth, particularly where influenced by the activities of beavers. Riparian environments usually supported healthy stands of western wheatgrass, the most productive species on the northern mixed prairie.[19] On 14 July 1805 at the Great Falls of the Missouri Meriwether Lewis noted that "the grass and weeds in this bottom are about 2 feet high; which is a much greater hight than we have seen them elsewhere this season. . . . the grass in the plains is not more than 3 inches high."[20] Sedges (*Carex atherodes*), exceptionally palatable to bison, also grew in moist depressions and swamps.[21] As a result of the better growing conditions, valley flats on the northwestern plains were about as productive as the fescue prairie.[22]

The fescue grasses, like other tall grasses along the margins of the plains, sprouted as much as a month later than the grasses of the mixed prairies, making them poor forage during early spring. Rough fescue, like other grasses of the aspen parklands, responded poorly to heavy grazing during the growing season.[23] In the south and on the highlands like the Cypress Hills, however, the fescue grasslands are more diverse than in the north, and the additional species found in the foothills region made those grasslands more tolerant of grazing than parkland grasses.[24] On the northwestern plains (especially in the foothills) the fescue grasslands, dry mixed prairie, and moist mixed prairie were often located very near one another, so bison herds needed to move only a short distance between them. As adjacent grasslands greened early in spring and fescue grasses stayed dormant, bison moved onto the moist mixed prairie, leaving the fescue to complete most of its growth cycle relatively undisturbed. Fescue grasses grew robustly in June and July and over the course of a year were markedly more productive than other grasses on the northwestern plains. They produced twice as much forage per acre as the moist mixed prairie and two or three times as much as the dry mixed prairie.[25]

The land's carrying capacity for large ungulates generally increased toward the north on the Great Plains, but on the northwestern plains also increased toward the west and southwest. The highlands, more common on the northwestern than the northeastern plains, also supported more luxuriant grass growth because of the greater precipitation and cooler temperatures (resulting in less evapotranspiration). Freshwater springs along slope breaks in upland regions watered flora and fauna. Rough topography and the river valleys (relatively deeper and wider than valleys to the east) provided lush grass growth and shelter from winter storms for both bison and humans.[26]

The mixed prairies included both cool- and warm-season grasses. Because of the topography, various early-growing and late-maturing species were even more likely to grow near each other. Bison that fed on fescue grasses all winter often had to move only short distances to graze on the early spring growth of June grass or wheatgrass. Bison on the flatter eastern plains did not have this advantage. Other characteristics of the western grasses also sustained buffalo. Taller grasses generally consist of a higher proportion of cell wall and thus are less nutritious than shorter grasses. This helps to explain why ungulates like the pronghorn antelope (*Antilocapra americana*), which required higher-quality forage than bison, were common on the western plains. More importantly, western grasses tend to cure well in the summer or fall. Curing is a process in which grasses retain high levels of digestible carbohydrates and a portion of their crude protein after their growth period ends. Although the grasses of the tallgrass prairie grow abundantly, they cure poorly on the stem. Their nutrition drops rapidly after the first fall frosts bring the growing season to an end. In contrast many grasses of the northern mixed grass and fescue prairies cure well and remain erect, retaining a lot more carbohydrates and protein than eastern grasses even until the grasses resume growing in the spring.[27]

Unique characteristics of fescue grasses make them particularly good dormant-season forage. They retain high crude protein levels exceptionally well, contain a very low percentage of lignin (a structural substance related to cellulose), and have a very low degree of cross-linking in the lignin-cellulose complex that forms the cell walls.[28] Although early ranchers had no scientific explanations, they were quick to notice the superior quality of fescue grasses during winter. McEachran noted:

> We were informed, and have no reason to doubt it, that these grasses do not wither and die as they do in a more humid climate, but, owing no doubt, to the purity and dryness of the air, they cure on

> their roots and make excellent hay. They thus preserve all their nutritious qualities, and make excellent feed for winter, a fact which is proved by the fat condition of all stock wintered in that country.[29]

Although the northwestern grasses were less productive than those of the northeastern plains, they were more nutritious during the critical period of the year when the quality of forage was at its nadir. Thus the neck of the hourglass was wider on the northwestern plains than one might expect.

Not only was more abundant forage available on the northwestern plains than in most neighboring areas, particularly in winter and early spring, but average climatic conditions also helped to sustain bison populations there during the winters. The Rocky Mountains obstruct the easterly flow of mild and moist Pacific air, and as a result the climate of the entire northern plains is continental, with long, cold winters and short, warm summers. The mountains produce a rain shadow effect, with precipitation (in the form of both rain and snow) being lighter on the northwestern plains than on the northeastern plains.

The Rocky Mountains also engender foehn winds known locally as chinooks. These warm, dry winds have the greatest effect on regions closest to the mountains, including the grasslands that abut and even penetrate the mountains south of the Bow River. In many of these areas chinook winds are so strong and constant that they have affected the shape of landforms. When air passes over the mountains, especially during the winter, the heat released by condensing and freezing water often produces warmer air temperatures east of the Rocky Mountains than to the west. The chinook winds from the southwest keep the southwestern part of the region (the chinook belt) generally snow free and relatively warm for much of the winter (see figure 1.4). The present-day city of Lethbridge, Alberta, averages thirty-five days of temperatures above 5° C (40° F) during the coldest three months of the year (December, January, and February), while Winnipeg, at nearly the same latitude, averages fewer than two.[30]

On his first visit with the Piegans in the winter of 1787–88, David Thompson noted that "our guide also told us that as we approached these [Rocky] mountains of snow we should find the weather become milder; This we could not believe, but it was so and the month of November was full as mild as the month of October at the trading house we left to the

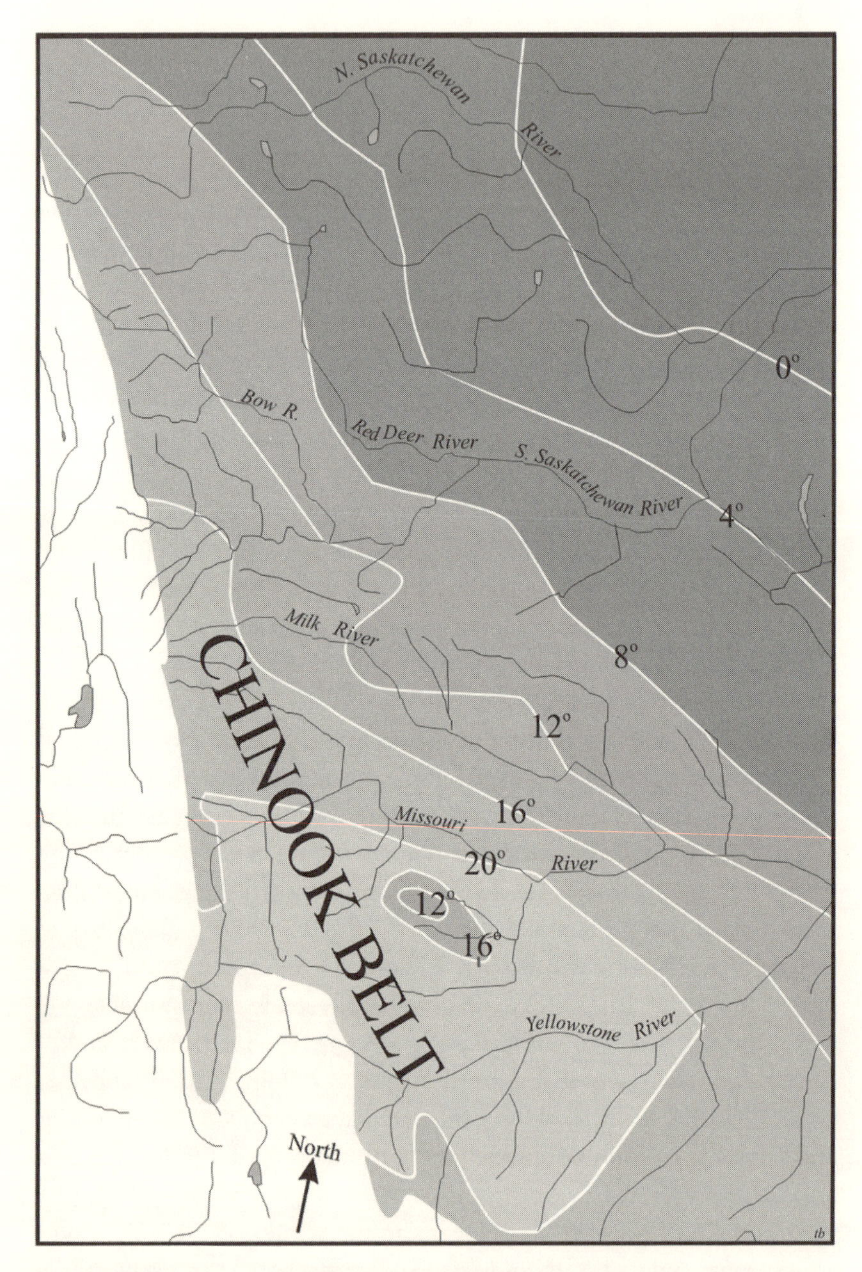

Fig. 1.4. Average January Temperatures on the Northwestern Plains (°F). Adapted from Longley, *Climate of the Prairie Provinces*; and Taylor, Edie, and Gritzner, *Montana in Maps*.

eastward. For the cold of these countries decreases as much by going westward as by going south."[31] A passage in Alexander Henry the Younger's journals illustrates the effects of a typical chinook. On the morning of 19 January 1811 at Rocky Mountain House the temperature was 17° F, the wind was out of the northwest, and there were two feet of snow on the ground. Then Henry heard a noise

> towards the Southward as if a strong gale of wind was passing. . . . At 12 OClock we began to observe the tops of Pines to bend to the wind, and in a short time the gale reached us from the S.W. Upon meeting with the face it was warm and gentle, but very soon increased to a [?] gale . . . at 1 Oclock the Thermometer stood at 56 above Zero, having rose 16 degrees in the course of 3/4 of an hour. . . . the wind increased in a most violent manner which continued during the night. . . . At sunrise the Thermometer stood at [?] above Zero, and to our great surprise we found the snow all melted away, and nothing but a few ponds of water remaining. Not one speck was to be seen in the plains, nor anywhere were the wind could reach it.[32]

In most years chinooks keep the forage easily accessible to grazing animals throughout the winter.[33] The drying winds helped cure the grasses, ensuring that the nutritive value of plants in this region remained higher than it was in moister regions of the plains.[34] A wedge of forest north of the Bow River separates the grasslands from the mountains, so the chinook winds affect the northern prairies less than they do the upper Missouri and South Saskatchewan River basins. Snow cover, although not usually deep, is more or less constant from November to March in the North Saskatchewan River basin. Above-freezing temperatures are considerably less frequent than they are in the south. Still, the North Saskatchewan basin has warmer average winter temperatures and no greater average depth of snow cover than the Red River region of the eastern prairies, which is farther south and lower in elevation.[35] The parkland region, characterized by plains rough fescue and wet meadows of luxuriant sedges, is more extensive in the extreme northwestern plains than anywhere else. Even there bison could find plentiful forage during the winter.

Biomes are not static ecological communities but dynamic systems. Certain forces that increased the productivity of the grasses of the northwestern plains in the past are now suppressed. Ample evidence from traders' journals, explorers' accounts, and the scientific literature shows that many areas along the margins of the northwestern plains became forested beginning in the 1870s and 1880s, when the effects of fire and bison were suppressed. Aridity is the primary ecological force in the development of the grassland biome, but lack of precipitation alone does not account for the extent of the grasslands of North America during the bison era. The prairie-forest ecotone (and the tall-grass prairie) was the product of climate, fire, and herbivore interaction. Fire formerly had an important effect on ecosystems throughout North America. It was important in preventing trees from invading areas along the margins of grasslands that had sufficient precipitation to support forests. The margins of the northern plains at the end of the bison era were more open than current conditions might suggest. The implications for the study of human history are important. Fire significantly enhanced the ability of the northwestern plains to support bison, because forests tended to invade the most productive fescue grasslands first, reducing the forage available to grazers.[36]

Natural and anthropogenic fires created and renewed grasslands in areas where the potential dominants were aspen or even spruce forests. Lightning ignited prairie fires both in the bison era and in the era of Euroamerican settlement.[37] Such fires killed most evergreen trees regardless of the time of year when they occurred and suppressed the growth of aspen and willows significantly. Aspens, however, are very fire hardy. Although fires can damage or kill above-ground growth, aspen and parkland shrubs like saskatoon (service) berries (*Amelanchier alnifolia*) responded by suckering profusely. Under these conditions the habits of bison and other ungulates were significant. Bison were primarily grazers, but they also browsed, trampled, and wallowed on young shoots of aspen.[38] Bison destroyed young trees when they congregated in thickets of aspen in the winter (when aspen shoots were brittle) to escape cold weather and in the summer to rub and scratch themselves.[39]

Shortly after arriving on the northeastern plains for the first time in August of 1800, Alexander Henry the Younger noted that

> the ravages of the Buffalos at this place is certainly astonishing. . . . The willows are intirely trampled and torn to atoms, even the bark

> of the smaller trees are in many places totally rub'd off by the Buffalo rubbing or scratching themselves against them. The Grass upon the first bank of the river is entirely worn away. The numerous paths (some of which are a foot deep in the hard turf) which comes out of the Plains to the bank of the River, and the vast quantity of dung which lays in every direction, gives this place the appearance of a civilized Country where Cattle have been kept for many years.[40]

Other ungulates common on the plains or parkland also suppressed the growth of forests. Mule deer (*Odocoileus hemionus*), moose (*Alces alces*), which prefer to browse rather than graze, and elk (*Cervus elaphus*), which will browse, suppressed tree growth.[41] Bison may have been as important as fire in suppressing the invasion of forests on the plains.[42]

There is ample evidence that cycles of drought combined with the effects of fire and bison suppressed tree growth on the margins of the plains during the bison era. Historical evidence and the speed with which aspen forests invaded the grasslands suggest that the parkland ecotone was significantly more open in the past. Copses of young aspens, even thickets of evergreens, existed as they do today in the parkland, but their growth and expansion were continually suppressed by fire and bison.

While fires and herbivores suppressed woody growth, they stimulated grass growth, which attracted grazers. Grasslands burned in fall, winter, or early spring begin growing a week or two earlier in the spring than unburned prairie.[43] Periodic burning and grazing also increases the overall productivity of most grasses (with the possible exception of short grasses). Ungulates gain weight more quickly when grazing on recently burned areas than on long-unburned grasses. Summer fires encouraged the grasslands to produce high-quality forage during a season when forage quality was generally declining. Herbivores were also enticed by the highly digestible and nutritious plains prickly pear cactus (*Opuntia polyacantha*) that had been singed.[44] The destruction of forage by fire was only a very short-term loss for foragers, more than compensated in the longer term.

Although scholars were slow to acknowledge the importance of the deliberate use of fire by indigenous peoples, Stephen Pyne noted that "the evidence for aboriginal burning in nearly every landscape of North America is so conclusive, and the consequences of fire suppression so visible, that it seems fantastic that a debate about whether Indians used broadcast fire or not should ever have taken place."[45] It was formerly well known that North American Indians commonly used fire. But when it became popular belief that fire was harmful to the environment and that good management meant the suppression of fire, the assumption that Indians were noble savages "in tune" with nature led to the denial that they had ever deliberately burned the grasslands. As ecologists and naturalists increasingly came to understand that fire is a natural disturbance that is not detrimental to the environment, scholars also became more willing to discuss the aboriginal use of fire. Today scholars know that the environment in the Americas from Tierra del Fuego to the Arctic Ocean and on other continents was "managed" by indigenous peoples for their own purposes before contact with the West.[46] American Indians had good reason to burn their environments: the majority of plant and animal species used by humans worldwide are most abundant in early development ecosystems, not mature or climax systems.[47]

Although there are no systematic historical studies on the use of fire by the indigenous peoples of the northern plains, several scholars have presented evidence that Indians deliberately set fire to the prairies.[48] Fires improved forage for large game like bison and elk but also created ideal conditions for fire-hardy saskatoon berries and prairie turnips. George Arthur has provided abundant evidence that the Indians of the northern plains understood that they could use fire to make their lives more secure. Lewis and Clark witnessed the Hidatsas burning the prairie near their villages in early March "for an early crop of grass, as an inducement for the buffalow to feed on."[49]

Wise managers could use fires strategically at different times of the year to influence the movement of bison herds. For example, they could burn areas of the northern mixed prairie in the autumn (or even winter) to direct the bison toward their wintering grounds and to encourage grasses in those areas to sprout earlier in the spring. They could burn fescue grasslands in the spring to drive the bison onto the plains and to encourage the fescue grasses to grow. In any season people might burn an area as they left it, so that grass and game would be more plentiful there when they returned. They could also burn fescue grasses in the fall to prevent the bison from wintering where they otherwise might. Traders frequently mentioned that Indians burned the prairie near trading posts to keep the bison out of their reach during the winter.[50] This practice forced the traders to buy their provisions from Indian bands during the winter, rather than hunting for themselves. The same strategy could be used in attempts to prevent herds from moving toward rival Indian groups.

Because of these topographical, vegetational, and climatic characteristics the northwestern plains supported exceptionally large bison herds, perhaps the largest on the Great Plains. This abundance itself was important to the indigenous peoples of the region, but the environment also made these bison particularly vulnerable to skilled hunters. Hunters could take advantage of the rolling terrain and the many rough breaks, particularly when they hunted communally. Bison jumps (in which bison are driven off a cliff) are a distinctive feature of the northwestern plains, rare elsewhere on the Great Plains but common in the upper Bow, Oldman, and upper Missouri River basins. A bison jump required undulating terrain that concealed at least a small bluff. There were many suitable sites on the northwestern plains. In rolling terrain insufficiently broken to allow for jumps, Indians constructed bison pounds in which bison were impounded. Bison jumps and pounds are especially significant because for many centuries indigenous hunters relied on them from late fall to early spring, the critical period of the year. The rugged terrain also made it easier for individual hunters to stalk bison.[51] Even the almost incessant winds helped hunters. Bison have a keen sense of smell and can be difficult to approach when the weather is calm or the winds very light and shifting. A consistent wind was particularly useful to communal hunters, who sometimes required several days to coax a bison herd to a kill site.

The groves of trees in deep valleys and other sheltered locations and the nearby forests provided shelter and important supplies of wood for human communities. It is little wonder that in 1802 Charles Le Raye heard the Crows extol their lands "as fertile beyond description and as inhabited by numerous bands of Indians. The buffalo, elk, cabree [pronghorn], deer, black and white bears are found there in vast multitudes."[52] The apparently austere Yellowstone basin hides a game paradise.

The exceptional bison habitat of the northwestern plains greatly influenced the region's human history. Most significantly, the northwestern plains attracted indigenous societies from all directions. Throughout the bison era the region was one of eager immigrants but reluctant emigrants.

Wherever they came from, and whatever subsistence strategies they may have used before, all migrant communities in the region quickly came to rely heavily upon the bison. Hunters had to follow the herds through extensive territories in their annual migration cycle. Although the region was sparsely populated, human communities in constant pursuit of the bison inevitably met one another frequently. If we hope to understand the history of human interaction, it is important to reconstruct the annual cycle of the bison herds and hunters on the northwestern plains.

CHAPTER TWO

THE ANNUAL CYCLE OF BISON AND HUNTERS

A few of [the Crows] assembled and draughted on a dressed skin, I believe a very good map of their Country, and the[y] showed me the place where at different seasons they were to be found.

They told me that in winter they were always to be found at a Park [bison pound or jump] by the foot of the Mountain a few miles from this or thereabouts. In the spring & Fall they are upon this [Yellowstone] River and in the summer upon the Tongue and Horses River.

—FRANÇOIS-ANTOINE LAROCQUE
1805

If nomadism is meant to imply wandering, very few human societies must ever have been nomadic. Nomadism instead suggests communities moving with the cycles of resource availability. Particular bands resorted seasonally to familiar locales where they had learned that their needs could be met. Typically hunting bands on the northwestern plains repeatedly visited familiar sites. Under adverse conditions they understood their environment well enough to make adjustments to their seasonal rounds. Observers have long agreed that the annual cycles of the hunting bands of the northwestern plains were closely tied to the annual cycle of the bison. Euroamerican traders remarked on this fact. Alexander Henry the Younger wrote of the Piegan bands that "it is the motion of the Buffalo which regulates their course throughout the year over this vast extent of meadow Country as they must always be near them to obtain a supply of provisions for their Families."[1] An understanding of the annual cycle of hunters and its influence on patterns of human interaction requires an awareness of bison habits.

Scholars have not agreed on past bison behavior, however. Because there are no more large, free-roaming herds, researchers cannot simply observe their normal behavior. Efforts to reconstruct past bison behavior have produced long-standing and lively debates. Until 1951 many believed that bison herds migrated en masse to the south for the winter. Frank Gilbert Roe's monumental study of the bison in 1951 put the theory of grand migration to rest. Roe argued that bison "wanderings were utterly erratic and unpredictable and might occur regardless of time, place, or season, with any number, in any direction, in any manner, under any conditions, and for any reason—which is to say, for no reason at all."[2] He made no systematic attempt to tie human movements to those of the bison; but based on his belief that "hunting peoples are basically what the characteristic habits of the predominating game species of their own tribal or hunting habitat force them to become," he contended that the unpredictability of the bison resulted in endless wandering and "the frequent historical occurrence of privation and even of positive famine in the lives of the Plains tribes."[3]

In 1962 Symmes C. Oliver published the first systematic attempt to explain plains Indian history in the context of the annual cycle of the bison.[4] He linked the remarkable similarities among plains Indian groups to the geographical and ecological characteristics of the region, attaching particular importance to the bison. Oliver's work set the stage for the scholarship of the 1970s and 1980s. Basing their studies on documentary records, archaeologists George Arthur and R. Grace Morgan and historical geographers D. W. Moodie and Arthur J. Ray argued that bison in the past migrated according to a seasonal pattern.[5]

Several archaeologists in the 1980s and 1990s, citing published historical sources, ecological knowledge, and archaeological evidence, questioned the seasonal migration thesis. In 1984 anthropologist Jeffery Hanson relied on modern bison behavior, historical documents, and grassland ecology to argue that bison were present in various grassland habitats all year round and migrated erratically only short distances in response to unpredictable natural and anthropogenic stimuli.[6] In 1996

Mary E. Malainey and Barbara L. Sherriff concluded that the traders' records contradicted the seasonal migration thesis. They used archaeological evidence and published primary sources to defend their theory that bison tended to winter on the open grasslands rather than in the parkland.[7]

Despite the skeptics, the seasonal migration thesis needs only minor modification. Traders' records, other historical documents, archaeological evidence, and scientific research show that bison concentrations varied seasonally according to regular patterns under normal conditions and in predictable ways under anomalous conditions. The regularity and predictability of bison movements greatly influenced human hunting societies and patterns of human interaction. Knowledgeable and experienced human hunters moved deliberately from place to place; they did not wander aimlessly looking for the bison or for their human neighbors. Under anomalous conditions, when movements of bison herds could depart significantly from regular patterns, hunting bands tapped the knowledge of their most experienced and skilled members to predict where the herds might be found or turned to alternative food sources. Those who depended on the bison learned their habits thoroughly. Even in the mid-nineteenth century Henry Youle Hind wrote that "the ranges of the buffalo in the north-western prairies are still maintained with great exactness, and old hunters, if the plains have not been burnt, can generally tell the direction in which herds will be found at certain seasons of the year. If the plains have been extensively burnt in the autumn, the search for the main herds during the following spring must depend on the course the fires have taken."[8]

The northwestern plains supported an awe-inspiring fauna. Herbivores included elk, deer, antelope, bighorn sheep, and ground squirrels; but the plains bison, as the dominant generalist foragers, were at the core of an elaborate ecosystem. Many predators, parasites, and scavengers depended largely on the bison. Packs of wolves patrolled the margins of the herds preying on wounded or sick animals or scavenging leftovers from a human hunt or other carcasses. Crows, ravens, and magpies searched constantly for abandoned carcasses, joined on the ground by a collection of coyotes and foxes. Grizzly and black bears subsisted mostly on roots and berries and ate only a modest amount of bison meat. Still, they were eager to make a meal of a dead bison, especially in spring, and a single grizzly could take a full-grown bison in ambush. Even humans were wise to give the plains grizzly a wide berth unless they were well armed. The bison also fed hordes of biting insects. Compared with other predators, human hunters were few but wise; their lives were intertwined with those of the bison.

Winter released its grip on the northern plains only grudgingly. Although heavy snowfalls and hard frosts were common even in April and May, particularly on the high plains and foothills, the first of the native cool-season grasses of the northern mixed prairie usually began growing in late March and early April. Snow persisted in the sheltered parklands, foothills, uplands, and valleys, and the fescue grasses remained dormant for several more weeks. The tender protein-rich cool-season grasses of the moist mixed prairie coaxed the bulk of the bison herds out of their sheltered winter range just as countless biting insects emerged to drive the herds into the open. George Bird Grinnell noted that "as spring opened, the buffalo would move down to the more flat prairie country away from the pis'kuns [buffalo jumps]."[9] Many of the cows led calves that had been born between early March and late June (bison can reproduce at a rate that approaches 18 percent per year), to replace the many bison lost to starvation, predators, and accidents during the previous year.[10] After subsisting on dried grasses since early autumn, the bison eagerly grazed the short swards of new growth. Immature cows and the bulls gained weight quickly, nursing cows more slowly.

The northwestern plains are dry, but May and June are the wettest months: cool northeasterly winds often meet with warmer air or with highlands to the south and west, producing rains in most years. When fed by these rains, the northern mixed prairie tolerated heavy bison grazing, producing new growth continuously as weather allowed. The fescue grasses, which responded poorly to spring and early summer grazing, were relatively undisturbed during their period of most robust growth between early May and early June.

Although the bison population peaked in spring, herds were widespread and highly mobile at that time of year as they grazed on the scanty new vegetation. With surface water at its peak availability, bison could use their entire range very effectively. Human hunters found it difficult to depend on the small and widely dispersed herds in spring. When hunters had a store of dried food, they could have remained in their sheltered winter camps, protected from the common April and May snow-

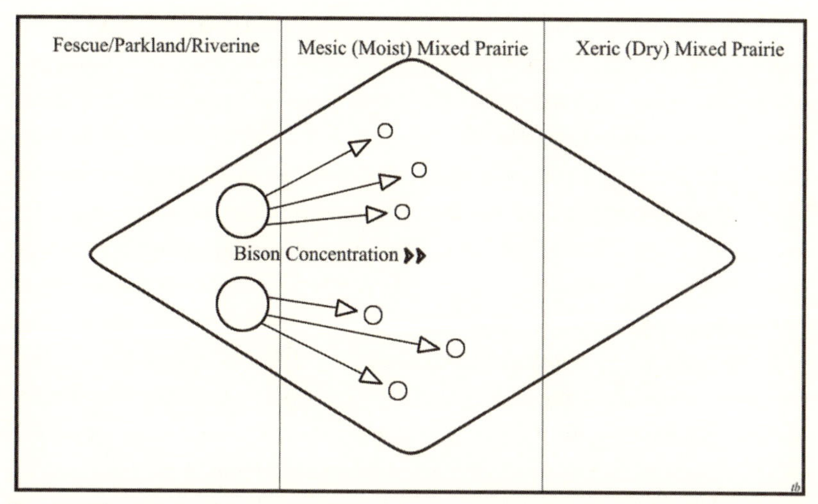

Fig. 2.1. Bison-Hunter Concentrations in Spring. Beginning in late March or early April bison were most concentrated on the moist mixed prairie, where cool-season grasses grew vigorously. Human hunting bands dispersed to follow and hunt the small mobile herds. Hunters took bison by decoying, stalking, and surrounding them. The diamond in the diagram represents the relative numbers of bison in the different areas of the prairie and the circles represent the relative size and movements of human bands.

storms, where they could prepare clothing for the summer months. They also turned to alternate species like elk, deer, and bighorn sheep or supplemented their diet with wildfowl eggs.[11] Bands could also gather or trade for bitterroot, prairie turnips, and camass bulbs in spring and early summer before their leaves withered and died, concealing their substantial roots.[12] These products were eaten immediately or dried and kept for later use. Needy groups also could turn to their kin for assistance.

During the spring most human communities, like the plains bison, gravitated toward the moist mixed prairie and dispersed into small mobile groups (see figure 2.1).[13] Even after they acquired guns hunters generally ran bison on horseback, killing them with bows and arrows. Bison could be taken individually with a bow and arrow or even with a gun without stampeding a herd, but this had to be done very carefully, especially with a gun. The hunter targeted the leading cow first and sought to kill each animal instantly to avoid stampeding the others. Although Indian hunters could take bison with guns, they did so only rarely before the 1860s.[14]

Exceptionally hot and dry weather in spring and early summer significantly reduced the carrying capacity of the prairies, particularly the summer pasture. The aridity made the grasses less productive, and the scarcity of water made part of the range inaccessible to herds. Two or more years of dry weather not only reduced productivity but began eliminating the more productive cool-season grasses from the drier portions of the dry mixed prairie. The altered constitution of the areas reduced grassland productivity and bison populations over a longer term. Unusually cool, wet weather had the opposite effect.

Bison and human communities were small and dispersed in spring. Hunters were oriented toward securing subsistence and preparing for the summer endeavors and expeditions. Bands were likely to meet neighboring bands with whom they had cooperative, reciprocal, and trading relationships, but they rarely launched war expeditions at this time.

As the weather grew warmer and drier in late spring and early summer, growth of the cool-season grasses slowed. By July, the warmest month on the northwestern plains, these grasses had flowered and their growth had nearly stopped, decreasing their palatability. From May to July, however, the warm-season grasses like blue grama, dominant in the dry mixed prairie, grew vigorously. Although these grasses were not very productive, they were protein-rich and appealing to bison. As the rut approached in early July, the bulls were sleek; this was the only time of the year when human hunters valued the meat of bulls as much as the meat of cows.[15]

When the bison came into rut in midsummer (peaking in early August on the northwestern plains), the mature bulls, which formed separate herds most of the year, mingled with the cow herds, establishing relatively few but enormous herds. Wherever they went, the herds left their mark. They stirred up dust and destroyed the vegetation with their wallowing and rubbing. Their dung and urine fouled the water and the air. The peculiar grunting of the excited males resounded over the landscape, night and day, like the roar of distant rapids.[16] The agitated bulls, warm weather, scanty forage, and biting insects kept the herds constantly on the move.[17]

By late summer, especially in dry years, many water sources in the dry prairie disappeared. As the weather cooled in August, the growth of warm-season grasses of the dry mixed prairie slowed and the forage became scanty. The bison again gravitated toward the moist mixed prairie and riverine habitats, where water sources and forage were more plentiful.

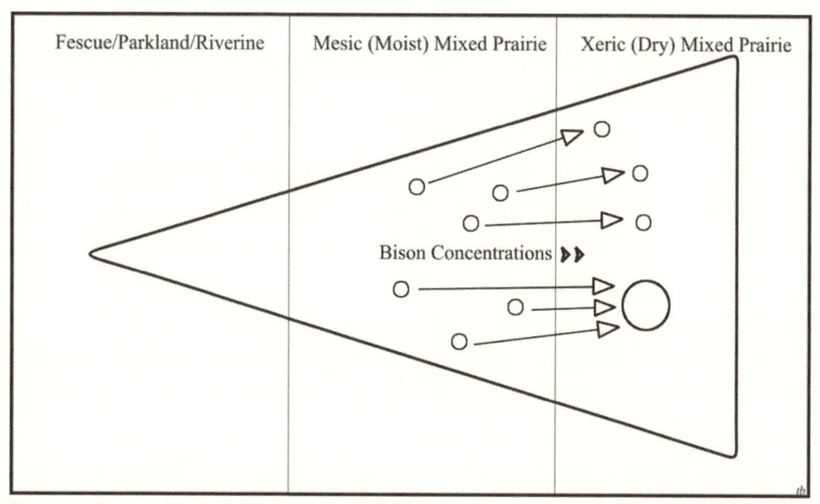

Fig. 2.2. Bison-Hunter Concentrations in Summer. By late June and early July bison were concentrated on the dry prairie, where the palatable warm-season grasses were growing vigorously. Pedestrian hunters may have been forced to stay in small mobile bands to follow the herds, although the formation of combined cow and bull herds in August may have allowed for larger encampments. Summer encampments were typically large in the equestrian era.

Grazing, summer fires, and falling temperatures could encourage cool-season grasses to send up new growth even in August.

Pedestrian and equestrian human settlement patterns during the summer may have differed considerably (see figure 2.2). During the equestrian era the largest human gatherings of the year took place in June or early July. Family and friends strengthened and renewed their relationships during the annual Sun Dance held at this time of the year. Summer was also a time for war and raiding expeditions. The gathering bison herds, the prolific saskatoon berries, and the long daylight hours encouraged large encampments.[18] Women gathered berries to be eaten fresh or dried and stored in bags for later use.[19] Men could travel considerable distances on horseback to find bison herds and transport meat.

Although bison might occasionally be driven into pounds or over jumps during the summer, they were most characteristically taken by running them on horseback with bows and arrows.[20] Anthony Henday noted in 1754 that the mounted hunters of the northwestern plains were "so expert that with one, or two, arrows they will drop a Buffalo."[21] The tendency of bison to herd together as they stampeded facilitated the hunt. Bison had greater endurance than even the fleetest horses, however, so the chase lasted only so long as the horses could keep up. Because unwise hunting methods could drive herds out of range of a camp, the large summer gatherings were governed by strong regulations regarding bison hunting and by police societies that enforced them. Leaders coordinated and planned hunts to make them as effective as possible.

Once the bison were rutting, it probably became more difficult to hunt them. Herds may have been larger, but they were less predictable and more mobile. Bulls, though less wary of danger, could be very aggressive and dangerous. Alexander Henry the Younger noted that bulls were "very fierce and not in the least timorous, as they will often turn upon a man and pursue him for some distance."[22] The largest human gatherings normally occurred early rather than late in the summer.

It is uncertain whether indigenous groups gathered in such large summer encampments during the pedestrian era. Archaeologists have found small summer camps but have not found evidence to show that large encampments were typical of pre-equestrian plains societies.[23] They may not have found large summer sites because it is harder to guess where they might have been located or because they are more likely to have been disturbed by modern agricultural activities. It is also possible that large summer encampments and Sun Dances were a post-horse phenomenon or became much more common during the equestrian era.

In equestrian days hunters could travel considerable distances to hunt the bison and to transport butchered meat, but in pedestrian days their possibilities were much more limited. Before the arrival of the horse summer hunts were conducted by surrounding herds or by stalking bison singly. Hunters lured animals by mimicking bison calves or approached them by camouflaging themselves as solitary wolves. They could take advantage of the bison's tendency to lie and rest during the warmest part of the day. Because bison have relatively poor eyesight but a keen sense of smell, hunters could camouflage themselves fairly easily, but they had to approach their quarry from downwind. Surrounding or stalking herds may have been no easier in the summer than it was in the spring, despite the existence of large herds. The level, treeless summer range of the bison afforded little cover in which to hide.[24] During the rut bulls were inattentive to hunters but were very dangerous when attacked and wounded.

The horse was particularly important to summer subsistence. The archaeological evidence of small summer encampments in the pedestrian era and lack of evidence of large summer gatherings, combined with our knowledge that the horse was especially suited to summer hunts, makes it plausible that summer encampments were typically much smaller in the pedestrian era than they were in the equestrian era.

Autumn is usually dry on the northwestern plains. Depending on forage and moisture conditions, herds could stay on the grasslands well into the fall. Skittish of ambushes by grizzlies, wolves, or humans, plains bison appear to have avoided the wooded country when possible. As the highly mobile herds wandered, they gravitated toward the fescue grasslands of the foothills, parkland, and deep river valleys. After their late start the highly productive fescue grasslands grew vigorously during the late spring and remained green into September and October.[25] By September they were the most nutritious grasses on the northern plains.[26] Cold weather and winter storms also forced bison herds to seek the thermal cover available in these habitats. Warm, snow-free weather can persist to the end of October in the north, however, and even longer in the south. While warm weather allowed the grasses to cure, it also encouraged bison herds to remain dispersed on the open plains. Then human hunters might set fire to the open grasslands to force herds to move toward their normal winter pastures.

According to George Grinnell, "in the last of the summer and early autumn, they [the Blackfoot bands] always had runners out, looking for the buffalo, to find where they were, and which way they were moving. In the early autumn, all the pis'kuns were repaired and strengthened, so as to be in good order for winter."[27] John Palliser noted that the Blackfoot bands turned at the end of summer from the open plains toward the foothills, where they expected the bison to seek shelter in the winter and where they themselves sought shelter and fuel (see figure 2.3).[28] In the broad valleys and in the hills the Indians gathered saskatoon berries, which had dried on the stem; they harvested choke cherries, which ripened just as their leaves cloaked the valley sides in brilliant red; or they gathered lodge poles from nearby evergreen forests and prepared winter clothing. Sometimes hunting parties resorted to the mountains to hunt bighorn sheep, whose horns were used for bowls and ladles.[29]

Hunters attempted communal bison hunts whenever circumstances permitted. If herds arrived at the rough country early, they were taken at

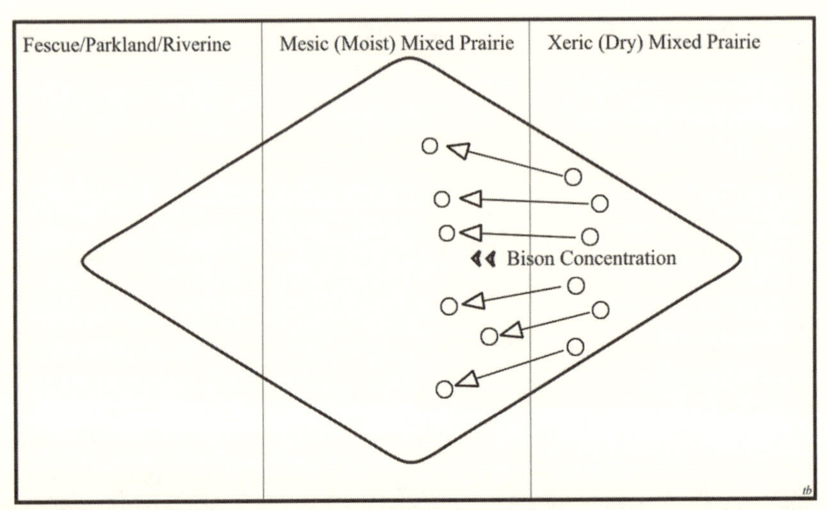

Fig. 2.3. Bison-Hunter Concentrations in Autumn. In August and September major concentrations of bison were on the moist prairie, where surface water supplies were still adequate and cool-season grasses were in their second growth cycle. At the end of September the combined cow and bull herds were separating again as the rut ended. Human hunting bands moved with the bison concentrations.

jump sites in autumn. Bands that wintered in the foothills relied on jumps. In the parkland and plains the hunters often depended on bison pounds, because the topography rarely allowed for jumps. William Pink provided what is probably the earliest description of a bison pound, but Matthew Cocking's journal provides a detailed early description of a Blackfoot pound:

> It is a Circle fenced round with trees laid one upon another at the foot of a Hill, seven feet high and a hundred yards in circumference, supported by letting trees into the Ground on each side and binding them with Leather thongs at top; the Entrance on the Hill side, where the Beasts can easily go over but when in cannot return, from this Entrance sticks are set up in form of a fence making an angle on each side extending from the Pound, beyond these to about one one-half mile distant Buffalo Dung or old roots are laid in heaps in the same direction as the fence, these are to frighten the Beasts from deviating on either side.[30]

Communal bison hunts of this type, however, occurred mostly in winter and early spring.[31] Hunters typically took bison in autumn by surrounding and stalking them in pre-horse days and by running them in the equestrian era. Mature bison bulls lost weight during the rut, but the cows and immature bison reached their prime in early autumn.[32] With the leanest period of the year coming, hunters sought to accumulate stores of preserved meat. In the warm, dry days of autumn butchered meat could easily be dried into a jerky or even processed into pemmican. Autumn, like spring, was a time of preparation for the coming season. Neighbors and friends probably visited and traded, but this was rarely the season to launch long-distance trading or war expeditions.

The crude protein content of grasses on the northern plains declined throughout the fall and winter, but the grasses of the northwestern plains retained their nutrition better than did the grasses farther east and were more abundant, particularly in the fescue belt, than the grasses farther south. Bison are better adapted to digest low-quality feed than cattle are and eat more grass to compensate for the poor forage.[33] Unlike cattle, they do not significantly increase their metabolism to generate more heat in cold weather. Instead they reduce their activity, conserving energy and forage.[34] Bison develop a much deeper and denser winter coat than cattle do and are particularly well adapted to winter conditions.[35]

In the chinook belt, where there is little winter snow accumulation, bison could graze even the remaining short, well-cured grasses of the northern mixed prairie, resorting to shelter only when the weather was cold and windy. The scanty snow, often blown into drifts, provided moisture that allowed the bison to graze areas inaccessible in the dry autumn period. Warm, snow-free winters often produced great hardship for human hunters, because the herds were so scattered that large bands could not depend on communal hunts and small bands could not kill enough animals to sustain their members.[36] Even when chinook conditions tempted hunters onto the plains, they must always have feared the sudden and unpredictable storms that could blow in from the north. The practice of burning grasslands in the fall and winter may have been part of an attempt to force animals that might have stayed on the open plains to move to their normal winter range.

In most winters the open plains were inhospitable to the bison, now in separate cow and bull herds. They were most numerous in areas where forage was plentiful and shelter nearby. The best forage and shelter could be found in the fescue crescent (the broad parkland belt near the North

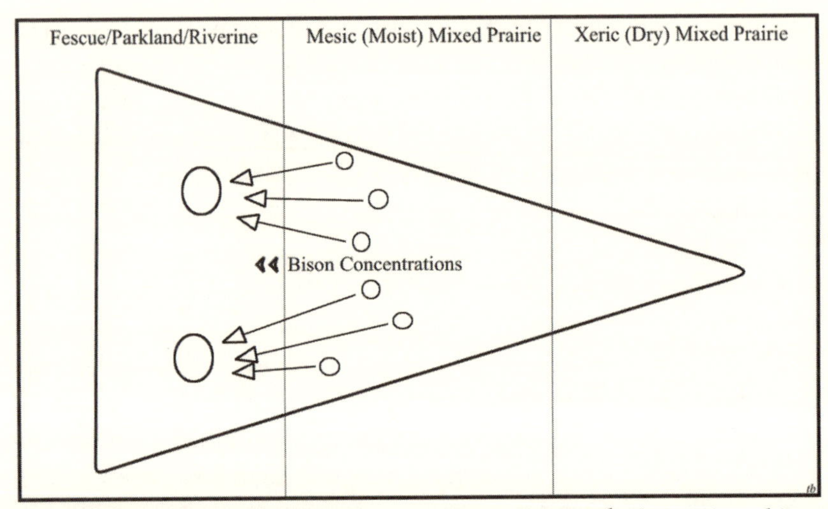

Fig. 2.4. Bison-Hunter Concentrations in Winter. During the long winter bison were concentrated in the fescue grasslands of the parkland and foothills and the sheltered riverine habitats. Pedestrian hunters often preceded the bison into these regions. Hunters could form large bands because the small and sedentary but concentrated separate cow and bull herds were vulnerable to attempts to trap them in pounds, jumps, and snow banks.

Saskatchewan River and the rough terrain of the foothills), in the uplands dotted throughout the region, and in the valleys of the major rivers. There were some very large herds of bison in winter throughout the northern plains.[37] In January of his first winter on the northern plains Alexander Henry the Younger noted: "I supposed I had seen incredible numbers of Buffalo in the fall coming up, but they were nothing in comparison to what I now beheld. The ground was perfectly covered in every direction of the compass, as far as the Eye could penetrate in the wood, and see over the plains, and every one in motion."[38] More often, however, bison gathered in many small herds.

According to Grinnell, "as winter drew near, the bison would again move up close to the mountains, and the Indians, as food began to become scarce, would follow them toward the pis'kuns."[39] Winter camps were located in the sheltered foothills and parklands (see figure 2.4). When choosing the sites for winter camps, hunters had to consider the availability of fuel for winter fires, forage for horses (in the equestrian era), shelter from winds and winter storms, and proximity to bison jumps or pounds. With most of the bison herds gathered in a relatively small area, humans could gather together as well. Large winter encampments were probably limited to the chinook belt. Smaller encampments were spaced a few miles apart in the river valleys (in keeping with the limited fuel and forage available in most locations).[40] Since forage for horse herds was not a consideration in pre-equestrian years, the advent of the horse may have encouraged this later settlement pattern.

The archaeological, historical, and ethnographic evidence shows that during both the pedestrian and equestrian eras Indians employed bison jumps and pounds more in the winter and early spring than at other times of the year.[41] In winter herds gathered in the broken terrain and valleys that lent themselves to pounds and jumps. Abundant and relatively small sedentary herds of cows and bulls were ideal for jumping or pounding. Hunters could maintain a camp near one jump or pound while scouts located cow herds and began luring them toward the kill site. The small, relatively inactive and predictable herds were more manageable than the larger, more active summer herds.[42] Most of the drive process involved carefully enticing a herd toward the chosen kill site (only the final, climactic episode required a stampede), so the habits of winter herds were more suitable for bison drives.

Horses were less useful in winter than in summer, being poorly adapted to strenuous work during cold weather. When it was cold and snowy Indians of the northwestern plains often used dogs to pack belongings, to reduce strain and minimize mortality among their horses. Even in warmer winters horses were less important to hunters than they were in the summer. Alexander Henry the Younger noted that "horses are sometimes made use of to collect and bring in the Buffalo [to pounds] but this manner never answers so effectually as people on foot, as it causes them to withdraw in a short time at a great distance. When horses are used the Buffalo are absolutely drove into the pound, whereas when the other method is pursued they are in a manner enticed to their own destruction."[43]

During most winters bison maintained themselves on the northwestern plains reasonably well, although their numbers declined. Mature animals lost weight gradually during the course of a winter as they drew on fat reserves to make up for the lack of quality forage. The surviving bison were lean by spring. Each animal's potential to provide sustenance for humans reached its lowest level in late March or early April. Some lean bison, many of them old bulls, were found all year round, but they were common in late winter and early spring. Early spring could be especially difficult outside the chinook belt, where the crusty snows of March made foraging difficult. John Palliser noted: "I have killed many fat buffaloes in the months of January and February; after which I have invariably found them lean, and sometimes seen the ground sprinkled with blood from the hardness of the surface, which the animal tries to shovel aside with its nose."[44]

Alexander Henry the Younger commented on the significant mortality among bison in the late winter and early spring on the northeastern plains. On 28 February 1801 he noted that "wolves and Crows are very numerous feeding on the many Buffalo carcasses that lay in every direction." A month later he remarked on the vast numbers of dead bison floating down the Red River. These animals had fallen through the thawing ice and were too weak to rescue themselves. Henry wrote: "I am informed that almost every spring it is the same, but not always in such immense numbers as this."[45] Long and severe winters caused the greatest bison mortality. After an unusually cold and snowy winter and late spring in 1788–89 William Tomison noted the great numbers of drowned bison along the banks of the Saskatchewan River near its forks.[46] In April 1810 Henry mentioned that a North West Company man at Fort Vermilion had visited the plains, where he "saw upwards of sixty dead Buffalo laying in the Plains, which generally is the case at this season of the year, when they are weak, and once they lay down, they cannot rise again."[47]

George Grinnell wrote that during the winter the Blackfoot bands killed "large numbers of buffalo, and would prepare great stores of dried meat."[48] Present knowledge of human metabolism explains why Indians on the northern plains tended to store dried meat and other foods in the fall and winter, even if they believed they could kill large numbers of animals throughout the year. Body fat in some game animals, including bison, can drop so low by late winter or early spring that their meat, if eaten indiscriminately, cannot sustain human life. Since protein-rich food stimulates metabolism, consumption of it actually increases caloric requirements and hastens starvation at a time when food is most scarce. In an effort to satisfy energy needs, the digestive system begins to convert proteins into carbohydrates, but this is an inefficient process that produces an accumulation of nitrogenous end products that poison the body, particularly when fluid intake is low.[49]

To avoid protein poisoning plains Indians had to maintain their intake of fat and carbohydrates during late winter and early spring. In some regions they did this by trading for high-carbohydrate agricultural products. For example, Cree and Assiniboine bands of the northeastern plains and the Crows of the northwestern plains acquired corn and other foods from the Mandans and Hidatsas. Many bands of the northwestern plains, however, had little or no access to trade in agricultural products. It was fortunate for them that aspects of the climate and vegetation in that area probably enabled bison to maintain greater stores of body fat than they could on the northeastern plains. The inhabitants of the northwestern plains did not rely as much on dried meat and pemmican as did groups farther east, but the Blackfoot hunters did make a special effort to kill bison cows in the early autumn, when their fat content was the highest, drying and storing this fat and meat to be consumed when the bison had become lean. They rendered most into jerky but processed some into pemmican, which contained a large amount of bison fat and grease together with dried berries. The Blackfoot bands also gathered, dried, and stored berries and roots like the prairie turnip for later use.

As winter wore on, hunters had to kill more and more bison to secure the same amount of nutrition. They ignored lean animals and discarded lean cuts of meat. Archaeological and documentary evidence shows that hunters commonly abandoned entire animals or harvested only the tongue, backfat, and fetus. Archaeological investigations of bison jumps and the documentary evidence indicate that hunters preferred to harvest and exploit cows, but especially fetuses and perhaps young calves.[50]

While visiting a Siksika bison pound near the Vermilion River in December 1809, Alexander Henry the Younger noted that, of the mangled carcasses in the pound, "the Bulls were mostly all entire, and none but good Cows were cut up."[51] HBC employees also learned to be selective. Thomas Stayner noted that while hunting at Manchester House in February 1790 he and Thomas Spence stalked a herd of bison on their hands and knees through deep snow for half an hour, but "when we came within Gun Shot of the Buffalo [we] found the whole herd so poor [that they were] not worth firing at. We started them a considerable distance then returned home."[52]

Hunters sought cows because they were in better condition than bulls in the winter, but also because cows' fat distribution made their meat easier to exploit.[53] Henry noted that "the Cow Buffalo have frequently *depouille* [backfat] of this [two inches] thickness, and some have been known to have even three Inches, but this is a rare thing to be found. The common condition is from one to Two Inches. A bull seldom has any extraordinary back fat; their fat is principally to be found in the inside of the animal."[54] While wintering with Sakatow's Piegan band on 10 February 1793, Peter Fidler noted that his hosts were "remarkably fond of [calves;] even when not more than the size of a quart pot they eat them. The greater part of the Cows the Indians now kill is merely for nothing else but for the calf."[55] After especially severe winters and late springs the developing fetuses and the nonpregnant cows may well have been the only source of fatty fresh meat.[56] The Indians of the plains longed to see the first green growth of spring, and their understanding that freshly burned prairie sprouted earlier than unburned prairie must have induced them to ignite dormant grasses. The sight of verdant prairie and young bison calves renewed the spirits of winter-weary communities.

Winter brought diverse human communities together along the margins of the plains. Plains bands that moved toward the margins of the plains during winter encountered other bands moving south from the subarctic forests or east across the Rocky Mountains.[57] These other bands also wintered on the margins of the plains to exploit the plains bison there. All of them were familiar with the wood or mountain bison (*Bison bison athabascae*) and the plains bison (*Bison bison bison*), whose combined range extended well beyond the northwestern plains (see figure 2.5).[58] South of the Bow River, where grasslands penetrated the mountainous valleys, bison and bison jumps were common.[59] Although not plentiful, bison were also found still farther south on the Columbia Plateau west of the continental divide at least until the end of the pedestrian era.[60] Archaeologists have even found bison jumps west of the continental divide.[61]

During the winter plains bands met subarctic and transmountain bands along the margins of the plains. The different groups often profited from peaceful interaction that included trade, but they sometimes fought. Plains bands also coalesced during winter. With bison concentrated in a small area, plains bands could cooperate in communal hunts. During the equestrian era, and probably earlier, by gathering in relatively large

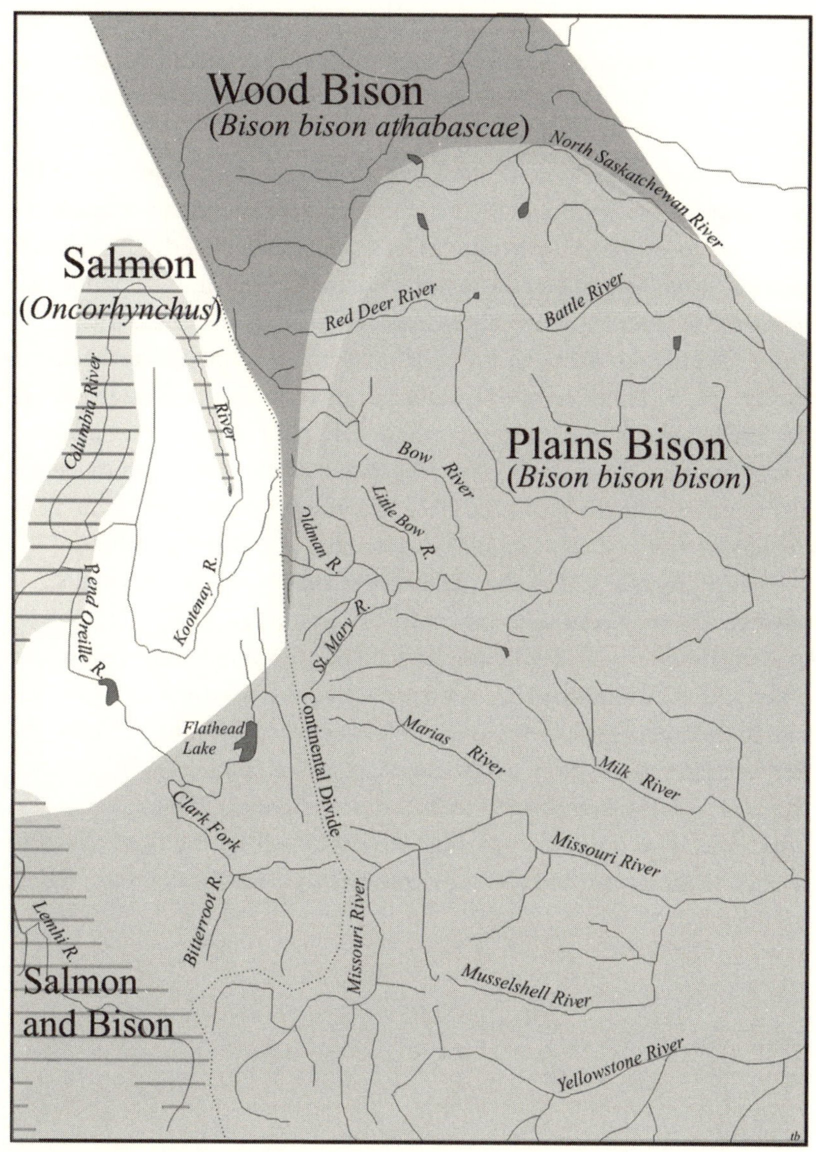

Fig. 2.5. Range of Bison and Salmon at the End of the Pedestrian Era. Adapted from Gates, Chowns, and Reynolds, "Wood Buffalo at the Crossroads," 140; and Northwest Power Planning Council, *Compilation*, chapter 3.

communities during the winter (while their neighbors outside the region were widely dispersed in small bands) plains bands could launch long-distance winter war and raiding excursions against their vulnerable enemies.

The northwestern plains were a fine environment for bison and for the humans who hunted them. Predictable variations in the concentration of the bison encouraged humans to follow similar seasonal rhythms. These annual cycles ensured that the region's relatively small human population frequently encountered neighbors from within and outside the plains and influenced human relations. Because of the bison's abundance, the northwestern plains bands enjoyed certain advantages over their neighbors. To the north, west, and southwest, where bison density was considerably lower, hunting bands were generally much smaller. Small bands generally found it difficult to displace large bands, so nature gave residents of the northwestern plains a military advantage over their neighbors.

The environment alone did not determine the course of human history, however. Humans did not, and could not, live in timeless harmony with their surroundings and their neighbors. They repeatedly seized opportunities and overcame challenges presented by the environment and their human neighbors. Over many centuries communities moved onto the northwestern plains from every direction. Some did so peacefully, some aggressively. They responded to the environment in different ways. It is important to understand not only the rhythms that the natural world seemed to encourage, but also the depth and dynamism of human history.

CHAPTER THREE

TRADE, WARFARE, AND DIPLOMACY FROM A.D. 200 TO THE EVE OF THE EQUESTRIAN ERA

Naw peu ooch eta cots from whence this river [Oldman River] Derives its name . . . is a place where Indians formerly assembled . . . to play at a particular game. . . . On my enquiring concerning the origin of this spot the Indians . . . said that a White man . . . came from the South many ages ago—& built this for the Indians to play at that is different nations whom he wished to meet here annually & bury all anamosities betwixt the different Tribes—by assembling here & playing together—they also say that this same person made the Buffalo—on purpose for the Indians, they describe him as a very old white headed man.

—PETER FIDLER
1792

The environment of the northwestern plains regulated human activity, but it left considerable latitude for human innovation. Although the abundance of the bison and the annual cycle of the herds influenced patterns of human interaction, these patterns were not timeless and unchanging. Through the ages distinct human communities have adapted to the environment of the northwestern plains differently. Relations between these peoples spanned the spectrum from peaceful and cooperative to hostile and competitive. The history of warfare, diplomacy, and trade is ancient. The archaeological record for the late pedestrian period, although fragmentary, is invaluable because it gives a sense of the depth, dynamism, and complexity of human history in the region.[1]

In the early part of this century many scholars, including prominent anthropologists such as Clark Wissler, Robert Lowie, and Ralph Linton, struggled to correct the image of the North American Indian as a bloodthirsty savage. Since then many anthropologists have defended romantic interpretations that downplayed the significance or even denied the evidence of warfare among indigenous societies before contact with Westerners. More recent scholars have shown that archaeological evidence undermines depictions of early plains warfare as a virtually bloodless prestige sport. They argue that warfare was important among indigenous societies before contact with the West and that Indians went to war to achieve practical and rational aims.[2] It is difficult to avoid the conclusion that warfare, including high-casualty wars and the mutilation of victims, long predates the arrival of Euroamericans.[3] Warfare on the Great Plains was a serious matter. By recognizing this we acknowledge the essential humanity of those who inhabited the region. The archaeological evidence helps us to see northwestern plains Indians not as stereotypical noble savages (or ignoble savages) but as sophisticated human beings who confronted difficult problems and tried to solve them thoughtfully and effectively.[4] When we romanticize communities we trivialize their history as much as we do when we demonize them.

While archaeological evidence establishes that warfare has long been a feature of life on the plains, it is too fragmentary for us to reconstruct trends over the years. The evidence suggests that the intensity and frequency of war were not continuous over time and space. Warfare was more common when communities were under stress.[5] It was probably more important among sedentary societies than among band communities and was generally less frequent where population densities were lower and communities smaller and more mobile. Warfare is dangerous and costly for any community, but its potential risks are greater and its likely rewards fewer among small nomadic communities than among larger sedentary ones. The temporary or permanent loss of a few adult males was far more significant for small bands. Small mobile bands could expect few rewards from war expeditions, especially during the pedestrian era. Target groups could be difficult to find. A war party might exact revenge, but even complete victory brought few other tangible gains, because mobile communities accumulated little surplus food and few material possessions. Similarly, victors did not realize much territorial gain or increased access to resources.

The evidence suggests that warfare was less frequent on the northwestern plains during the pedestrian era than it was in areas where human settlements were larger and more sedentary and during the subsequent equestrian era. In the early part of the twentieth century Flathead informants told James Teit that the arrival of the horse had greatly changed the frequency of warfare on the northwestern plains. Before the arrival of horses, "peace generally prevailed among the various tribes and there was no continual warfare like that which developed after the introduction of the horse and the migrations of eastern tribes westward and of Blackfoot tribes southward."[6] The archaeological evidence that small summer camps were common in the pedestrian era reinforces the impression that warfare was less significant then. This evidence is not unambiguous, however. In both the pedestrian and equestrian eras dire necessity sometimes forced hunting communities to disperse, even at the risk of exposing themselves to danger of attack by a larger party.

The argument that warfare became more frequent in the equestrian era should not obscure the fact that it played an important role in the late pedestrian era. Documentary and archaeological evidence confirms this. In the 1780s Young Man (Saukamappee), an elderly Cree-born Piegan leader, recalled a particular battle that took place near the Eagle Hills around 1730. The Cree participants in the battle already had access to iron, but in other respects the battle was, Young Man said, typical of warfare before the arrival of horse and gun. According to Thompson's recollection of Young Man's story, spies of a Piegan-Cree war party discovered a large Shoshoni camp near the Eagle Hills and prepared for battle:

> we had to cross the River in canoes, and on rafts, which we carefully secured for our retreat. When we had crossed and numbered our men, we were about 350 warriors . . . they had their scouts out, and came to meet us. Both parties made a great show of their numbers, and I thought that they were more numerous than ourselves.
>
> After some singing and dancing, they sat down on the ground, and placed their large shields before them, which covered them: We did the same, but our shields were not so many, and some of our

> shields had to shelter two men. Theirs were all placed touching each other; their Bows were not so long as ours, but of better wood, and the back covered with the sinews of the Bisons which made them very elastic, and their arrows went a long way and whizzed about us as balls do from guns. They were all headed with a sharp smooth, black stone (flint) which broke when it struck anything. Our iron headed arrows did not go through their shields but stuck in them; On both sides several were wounded, but none lay on the ground; and night put an end to the battle, without a scalp being taken on either side, and in those days such was the result, unless one party was more numerous than the other. The great mischief of war then, was as now, by attacking and destroying small camps of ten to thirty tents, which are obliged to separate for hunting.[7]

Young Man's account implies that war became more frequent and deadly in the equestrian era, but it also emphasizes the continuity in warfare from the late pedestrian to the early equestrian era. Rock art reinforces his portrayal of war in the pedestrian era. At many locations on the northwestern plains depictions of shield-bearing pedestrian warriors reveal that warfare was an important aspect of life before the arrival of the horse (see figure 3.1). This shield-bearing pedestrian warrior is by far the most common, and perhaps the oldest, motif in the over ninety rock art sites on the high plains of Montana and southern Alberta.[8] J. D. Keyser has argued that the pictographs and petroglyphs at the Writing-on-Stone site along the Milk River suggest that war honors were less important for the acquisition of status in society during the pedestrian era than they were to become in the equestrian era.[9] Stuart and Betty Lu Conner have suggested that the depictions of pedestrian warriors were probably autobiographical and that some portrayals of pedestrian warriors included coup sticks.[10] Warfare in the pedestrian era had social and ritual as well as military and economic importance.

Douglas Bamforth has generalized that "cultural-ecological research in anthropology makes it clear that humans rarely engage in extremely expensive patterns of behavior without very good reason."[11] If parties could resort to less dangerous forms of aggression or alternate forms of dispute resolution, they generally did. For example, one party might achieve certain aims yet avoid a direct military conflict by threatening and

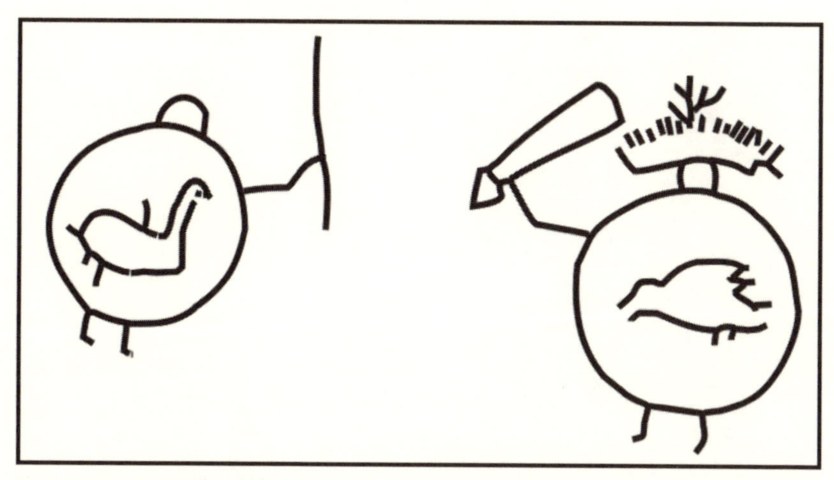

Fig. 3.1. Petroglyph of Shield-Bearing Warriors in Hand-to-Hand Combat. This petroglyph from Writing-on-Stone near the Sweetgrass Hills shows two pedestrian shield-bearing warriors in combat, using a lance and a club. Redrawn from Keyser, "Writing-on-Stone," 68.

intimidating another or by burning the prairie near a rival camp. As the epigraph to this chapter indicates, sometimes enemy groups settled scores by gaming together. Alternately parties might exchange gifts and smoke tobacco together to atone for offenses and restore or establish cordial relations. At present the archaeological evidence for these kinds of interactions on the northern plains is slight and difficult to interpret. Westerners recorded evidence of the aggressive use of fire and of interethnic gambling almost as soon as they encountered indigenous groups on the northwestern plains, however, which implies that practices like these were well established by the late pedestrian era.

Communities interacted not only competitively but also cooperatively. Trade, including long-distance trade, has an ancient history on the northwestern plains and became increasingly elaborate about two or three thousand years ago.[12] Much of the trade may have been in perishables and thus is archeologically invisible; but the surviving evidence proves that residents of the northwestern plains traded for many resources and goods, including apparent luxury items, that were not available in the region. For example, dentalium shells from the Pacific coast were imported to the northwestern plains hundreds of years before the arrival of the

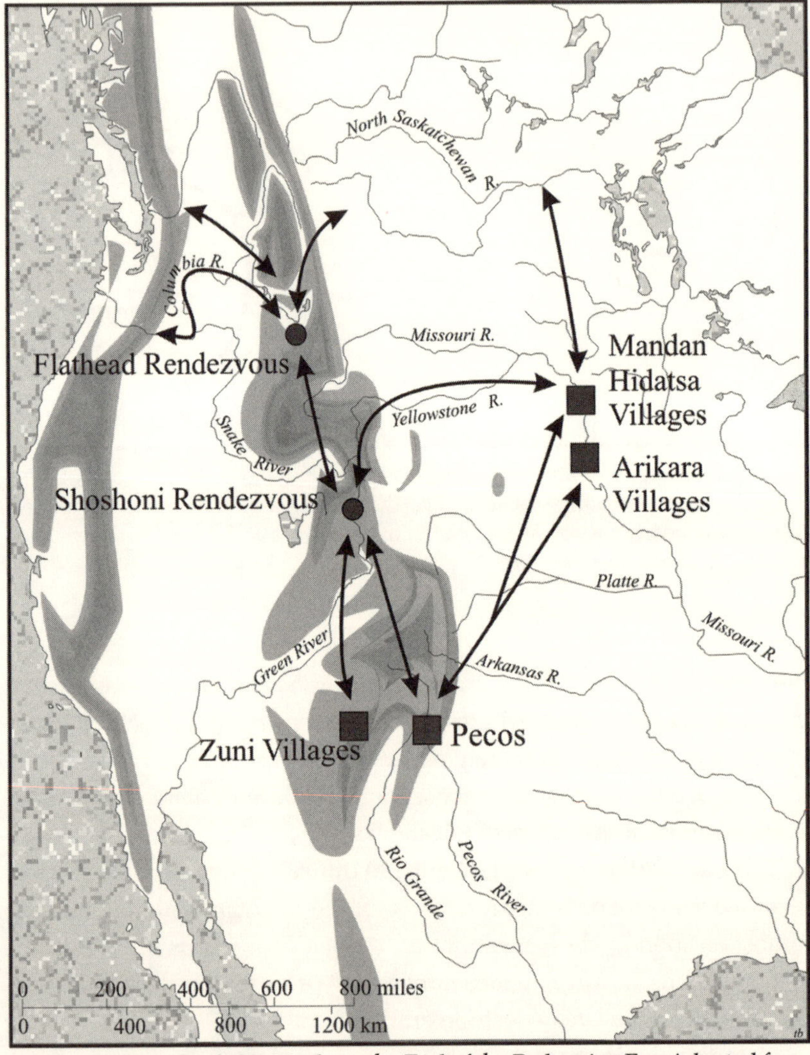

Fig. 3.2. Primary Trade Networks at the End of the Pedestrian Era. Adapted from Swagerty, "Indian Trade," 352.

horse.[13] Many of the trade networks of the late pedestrian era (see figure 3.2) remained important after the arrival of the horse.

The same range of human interactions marked the pedestrian and the equestrian eras on the northwestern plains. The arrival of horses, Western goods, and Euroamericans themselves, while very significant, did not initiate long-distance trade or warfare. Trade, warfare, and diplomacy were important to pedestrian plains societies, although it is difficult to trace and explain the changing patterns in human interaction.

Not long ago scholars assumed that the indigenous history of the northwestern plains was both recent and static. Archaeologists have now dispelled these notions. We know that hunting communities have occupied the northwestern plains since shortly after the end of the last ice age. Humans already inhabited the area by about ten thousand years ago, when it became a region of grasslands. The borders and constitution of these grasslands continued to change over the next millennia, but the region acquired many of its unique environmental characteristics thousands of years ago.

Although the limited archaeological evidence for the earliest millennia of human occupation makes it impossible to speculate on patterns of human interaction, it does show that human societies became increasingly sophisticated in their ability to subsist in the region. They began using bison jumps at least 5,700 years ago.[14] In the southern plains climate shifts periodically made communal bison hunting impossible, but Indians on the northwestern plains developed increasingly effective subsistence strategies over the millennia. About 4,800 years ago they began producing pemmican, which became an important food source during the leanest periods of the year.[15] Evidence from the Head-Smashed-In Buffalo Jump suggests that pemmican production increased dramatically at the beginning of the late pedestrian era.[16] By easing the year-round reliance on bison meat and facilitating trade in meat products, pemmican allowed the development of large, semisedentary winter encampments consisting of at least fifty tents.[17]

One of the most fascinating puzzles in the history of the northwestern plains during the pedestrian era relates to the arrival of the bow and arrow. Most archaeologists agree that bow and arrow technology first appeared on the northwestern plains between A.D. 150 and 250. Until that time hunters had used either hand-propelled spears or atlatls—dart throwers that permitted hunters to propel darts much further and with more power than they could throw spears (see figure 3.3). According to many archaeologists, the coming of the bow and arrow marked a turning point that ushered in the era of the classic pedestrian plains Indian way of life.[18] It coincided with a dramatic increase in the population of the

Fig. 3.3. The Atlatl in Use. The atlatl was a powerful tool that effectively lengthened the human arm, enabling hunters to throw darts much farther. Although simple in appearance, effective atlatls were difficult to make. Because atlatl technology and bow and arrow technology each had advantages, the bow and arrow did not rapidly replace the atlatl when it was introduced on the northwestern plains. Redrawn from a sketch by Knut R. Fladmark in the University of British Columbia Museum of Anthropology Collection.

northwestern plains. Thomas Kehoe has argued that the "complex, ritualized, patterned" bison drives date from this period.[19] Both the number and size of human camps grew.[20]

The arrival of the bow and arrow inevitably affected patterns of human interaction on the northwestern plains, but the archaeological evidence allows few firm conclusions about how these changes occurred. We can be sure that groups who were accustomed to using atlatls did not quickly abandon them and embrace the bow and arrow technology. Neither did groups who possessed bow and arrow technology simply displace the prior inhabitants. The transition from atlatl to bow and arrow technology was a slow, complex, and uneven process. The bow and arrow and dart thrower were evidently each well suited to subsistence in specific environmental contexts.

Apparently an archaeological culture known as Avonlea introduced bow and arrow technology to the northwestern plains.[21] The oldest evidence of these people is from the dry mixed prairie zone east of the Cypress Hills, although archaeologists question the notion that the group developed in that location.[22] The use of bow and arrow, the presence of dentalium shells, and the commonness of obsidian in Avonlea sites suggest that they arrived from the west and maintained trading relations with people there. Evidence regarding the arrival of the bow and arrow in North America reinforces this conclusion. Inuit migrants apparently introduced the bow and arrow to the continent from the northwest. The technology reached the Fraser Canyon by 350 B.C. and spread east into the interior in the early centuries A.D.[23] It seems that the technology arrived on the northwestern plains before it arrived on the northern periphery of the plains, the Great Basin, the Columbia plateau, and the eastern plains.[24] The oldest archaeological manifestations of bow and arrow technology on the northwestern plains are not along their western margins, however, but on the dry mixed prairie. If the technology did "leapfrog" from the Rocky Mountains to the dry core of the plains, environmental factors may help explain why.

In A.D. 200 societies of a culture that archaeologists have named the Besant phase dominated much of the northern plains.[25] Then they encountered people using bows and arrows. Had the bow and arrow been much superior to the dart thrower as a subsistence tool or military weapon, the atlatl-hunting Besant groups would almost certainly have taken up the superior technology or have been displaced by or integrated into the newcomers. Instead the Avonlea groups first occupied the drier regions of the northern mixed prairie, leaving the apparently more populous Besant communities to exploit the most abundant bison herds along the margins of the plains. Over several hundred years the Avonlea groups gradually expanded until they dominated the entire northwestern plains, from the North Saskatchewan River to the Yellowstone River basin.[26] Yet for centuries two groups of communities, with different subsistence technologies (and a different material culture), coexisted on the northwestern plains: the atlatl hunters apparently spent winters in the foothills and summers on the prairie, while the Avonlea communities wintered in the parklands and the valleys and moved to the prairies in the summer.

This long coexistence of groups depending on the bow and arrow and groups depending on the dart thrower suggests that each technology was well suited to particular environments and situations. The atlatl seems ideal for the uneven terrain and large bison populations of the more westerly areas as a powerful subsistence tool. The relatively large and heavy dart penetrated bones and flesh, causing massive and devastating injury even when hurled from a considerable distance.[27] Groups using sophisticated meat-preservation methods and occupying regions where bison density and topographic relief made bison jumps particularly reliable may have found no reason to abandon the atlatl as a weapon. In other areas groups may have found the bow and arrow more suited to their needs.

The Avonlea phase slowly expanded from a core area on the relatively dry and flat portions of the mixed grasslands. The Avonlea people were as sophisticated at communal bison hunting as Besant hunters were; at one site near the South Saskatchewan River Avonlea people built remarkable earthen and stone walls to impound bison.[28] Their effectiveness in small-scale hunts may have been the key to their success, however. There were far fewer potential communal kill sites on the flatter terrain at the core of the plains than at the margins, and bison pounds required greater effort to prepare and maintain than bison jumps did. When hunting communally, Besant peoples relied largely on bison jumps, Avonlea hunters on bison pounds. More importantly, although Avonlea hunters were consummate communal hunters, they could survive only if they could also hunt effectively as individuals and in small groups on the flat and sparse grasslands. In contrast to a hunter with a dart thrower, an archer can draw and release the string of a bow slowly and precisely while remaining hidden and relatively still, without spooking the herd. The bow and arrow, a weapon of stealth and accuracy, seems to be the superior subsistence tool on the flatter, drier plains.

The Avonlea hunters also had a broader diet than Besant hunters. They were far more likely to eat fish.[29] They were also very skilled communal hunters of pronghorn antelope.[30] This may help to explain their success on the drier portions of the northwestern plains when they had little access to the fescue grasslands in summer or winter.

Although archaeologists do not know much about the relationships between the Avonlea and Besant communities, there is no evidence that members of the two groups ever camped together. The material cultures did not merge, suggesting that they rarely intermarried or mingled. It seems unlikely that the two waged incessant warfare, for the two communities shared much of the northwestern plains for hundreds of years. The Avonlea peoples gradually spread outward from their initial core area between A.D. 200 and 1000. By about A.D. 800 Avonlea people left artifacts in the Kootenay basin west of the Rockies.[31] Perhaps the dry Scandic climate episode from about A.D. 250 to 850 reduced bison populations and thus decreased the opportunity for communal bison hunts. Under the drier conditions the bow-hunting Avonlea peoples with a broader subsistence base probably enjoyed a growing advantage over atlatl hunters.[32]

Ironically, after they adopted the bow and arrow the descendants of the Besant peoples may have eclipsed those who originally brought the technology to the plains. Most archaeologists believe that the waning of the Besant phase and the emergence of the Old Women's phase between A.D. 750 and 850 represent not the invasion of a new society but members of the Besant groups who adopted the bow and arrow.[33] If this is so, the Besant groups on the drier and flatter parts of the northern plains appear to have adopted the bow and arrow, which was especially well suited to these areas, before those in the foothills. The Old Women's arrow tips were crude compared to the skillfully and consistently crafted Avonlea points; but after the Old Women's phase developed the Avonlea phase began to wane, disappearing from the western plains around 1150. The archaeological evidence gives few clues as to what happened. Old Women's communities may have gradually eliminated or integrated Avonlea communities. Perhaps the Avonlea people were driven southward and westward. A combination of these events may have occurred, with different bands following different paths. Some have argued that similarities between Old Women's ceramics and Avonlea ceramics suggest that members of the Avonlea societies contributed to the evolution of Old Women's material culture.[34]

The Old Women's phase, expressed in several subphases, expanded over most of the northwestern plains between 1200 and 1750. Metal projectile points and glass beads found in a late Old Women's phase site near the South Saskatchewan River in southeastern Alberta tell us that this phase persisted into the era when Euroamerican traders arrived.[35] By that time, however, other groups with effective subsistence strategies connected with horticulture also influenced life in the region.

About five thousand years ago Indians of the Tehuacán Valley near present-day Mexico City began cultivating corn. After a long history there, corn cultivation spread into southwestern and eastern North America over many centuries. On the northern plains, in what some describe as one of the greatest achievements in plant breeding, aboriginal societies "transformed corn from a warm-weather plant which required high daytime and nighttime temperatures during a growing season of 150 days or more, to a tough, compact plant that matured in 60 days and resisted drought, wind, cool temperatures, and even frost."[36]

Corn agriculture was central to the development of Mississippian cultures of the lower Mississippi Valley after A.D. 800. Foremost among these settlements was a large urban cluster near the confluence of the Missouri and Mississippi Rivers known to archaeologists as Cahokia. Between 900 and 1300 Cahokia became the largest political chiefdom in the history of North America,[37] reaching its apogee around 1050. This community of about 30,000 people dominated an immense trade and transportation network oriented toward the north and northwest.[38] Mississippian influences evidently extended even to the northwestern plains, although to an unknown extent.

Shortly after A.D. 1100, when Cahokia was at its zenith, certain communities, perhaps refugees, began settling in horticultural villages in the middle Missouri basin near the Knife and Heart Rivers.[39] Using bison shoulder bones as hoes, they cultivated the easily worked alluvial soils and planted a variety of corn distinct from that grown on the Mississippi bottomlands but perfectly suited to conditions on the middle Missouri. Northern flint corn resisted rotting in the cold, wet soil of spring and tolerated summer drought well, permitting the development of viable agricultural villages in scattered locations along the Missouri River near large bison herds. The self-sufficient villages on the middle Missouri may have contributed to the decline of Cahokia by 1200.[40]

Although they were smaller, the middle Missouri villages influenced the history of the northwestern plains after 1200 more than the Mississippians did before. The middle Missouri villages were diverse. Their founders were Siouan-speaking ancestors of the Mandans and Hidatsas. The linguistic differences among Siouan languages suggest that these ancestors left the proto-Siouan homeland before the time of Christ. Differences between Hidatsa dialects and Mandan also indicate that these two Siouan groups were separated for many centuries.[41] Their paths converged again when they settled in the middle Missouri villages.

It seems that ancestors of the Awatixa Hidatsas, evidently the first Hidatsas to arrive, were later joined by forebears of the Awaxawi Hidatsas and the "Hidatsas proper" or "Willow Indians," who may have arrived as late as the 1600s.[42] By that time the ancestors of the Mandans had also joined them. Throughout the next centuries villagers integrated members of various other communities into their society, and the Middle Missouri Tradition became what one archaeologist has described as a "cultural melting pot."[43] "Cultural mosaic" might be a more accurate term, for despite their proximity over a long period, the villagers maintained their distinct languages and dialects even as they left Middle Missouri Tradition artifacts that are indistinguishable from one another. The three Hidatsa groups themselves settled in three groups of villages that maintained distinct, mutually intelligible dialects despite their physical proximity and constant interaction.[44]

Indians north of the Knife and Heart Rivers never adopted substantial agriculture. Between about 1150 and 1200 some members of the Missouri villages apparently introduced horticulture to the Souris, Assiniboine, and Red River valleys to the north, where some other local groups like the Blackduck and Duck Bay peoples apparently joined them. Other migrants from the south, beginning around the year 1000 and intensifying with the decline and collapse of Cahokia and the Mississippian centers around A.D. 1300, added to the diversity of societies in the Red and Souris River basin.[45] Yet communities north of the Missouri seem to have abandoned agriculture after 1500.[46] The advent of horticulture on the northern plains came during a relatively warm and wet period known as the neo-Atlantic episode (approximately A.D. 850–1250). Perhaps the onset of the drier and cooler Pacific climate episode (A.D. 1250–1550) made substantial corn agriculture in this region unfeasible, as it did in other places on the plains.

After indigenous groups north of the Missouri River abandoned horticulture, the middle Missouri villagers again became the northernmost horticulturalists on the northern plains. This gave them opportunities and challenges similar to those experienced by the Hurons, the northernmost horticulturalists in the Great Lakes region. The middle Missouri villagers, in contrast to agriculturalists on the central plains (Pawnees, Omahas, Poncas, and Otos), developed an extensive trading network oriented toward the north. Hunting societies were their natural trading partners. The northernmost horticulturalists were most likely to suffer crop failure and local resource depletion, however. Poor harvests and a shortage of timber forced villagers to move their villages periodically; more importantly, the middle Missouri villagers always combined their horticultural efforts with hunting and gathering, especially when their crops failed.[47] The villagers never abandoned the hunt, and the drier and cooler climate that prevailed after 1250 made horticulture more tenuous and bison hunting relatively more rewarding.

The climate changes of the thirteenth century appear to have strained communities in many areas of the Great Plains. Farther south some communities abandoned their villages; others fortified theirs as population pressures led to warfare.[48] Between 1300 and 1500 the middle Missouri villagers attempted to settle new sites in more favorable locations downstream but were prevented from doing so by Caddoan-speaking members of the Coalescent Tradition, probably ancestral to the modern Arikaras, who occupied villages to the south. As part of an ongoing struggle for agricultural land, it may have been the middle Missouri villagers who killed and mutilated about five hundred villagers at Crow Creek around 1325.[49]

The archaeological evidence suggests that when rivals thwarted their efforts to move onto more favorable agricultural land the Hidatsas increasingly combined their horticultural efforts with substantial reliance on hunting and gathering and on trade with other hunting and gathering societies on the northern plains. The people who left artifacts that archaeologists connect with the Mortlach phase were crucial to the trading network centered on the Missouri villages. Between 1300 and 1750 the Mortlach people expanded onto much of the northern plains and drove Old Women's communities west out of the South Saskatchewan River region.[50] They apparently hunted bison on the open prairies in summer and primarily in riverine environments in the winter, although northern groups may have wintered in the parkland. Mortlach sites are very large and often contain intensively processed bison bones, perhaps reflecting efforts to produce preserved meats that could be brought to the agricultural villages to the south or eaten during the long, cold winters.[51]

The variation among Mortlach artifacts betrays these peoples' involvement in trade and other kinds of peaceful exchange. The northern Mortlach peoples, who were more oriented toward the parklands, frequently visited and traded with the Selkirk peoples, who were expanding westward through the forests and parkland from the Saskatchewan River.[52] South of the parkland Mortlach artifacts include pottery that is clearly connected to the middle Missouri villages. Mortlach communities evidently acted as a conduit for the movement of people, goods, and ideas between the mobile hunting societies and the semisedentary horticultural populations on the northern plains. Their presence on the northern plains, including parts of the northwestern plains, reached its maximum extent in the late pedestrian era. Some of the late Mortlach sites contain European trade goods that these people probably acquired from the Crees.[53]

The middle Missouri villagers' influence on the northwestern plains continued to grow after 1600. While the Mortlach groups were affiliated with the middle Missouri villages, people connected with the One Gun phase were almost certainly actual migrants from the middle Missouri groups onto the northwestern plains. Whereas the Mortlach phase is typical of the northeastern and north-central plains, the One Gun phase is typical of the northwestern plains. The One Gun phase includes middle Missouri–style fortified village plans far removed from the actual middle Missouri villages. Although the One Gun peoples erected earthlodges, there is as yet no evidence that they made any attempt at growing corn. The One Gun people were probably nomadic hunters who erected fortified villages with the intention of occupying them seasonally.

The Hagen site, on the lower Yellowstone River near the present site of Glendive, Montana, probably dates from between 1675 and 1700. It evidently combined mud houses and tipis. Archaeologist William Mulloy interpreted this, and the evidence of small-scale horticulture, as indicating a transition from sedentary horticulturalism to nomadic hunting. He recognized that this was primarily a bison hunting community, however.[54] Bison scapula hoes suggest that these people planted something (perhaps tobacco), but there is no evidence that it was corn. The number of sherds found at the Hagen site suggests that its occupation spanned more than one season.[55]

The One Gun peoples also established the Cluny earthlodge village near a convenient ford on the Bow River east of the present city of Calgary, Alberta, during the early equestrian era, probably around 1740.[56] This is the most northwesterly known fortified village and may have been made possible by the mobility afforded by the horse.[57] Characteristics of this fortified village, like those of the Hagen site, suggest that it was almost certainly not an attempt to establish a horticultural settlement. There is no evidence that the builders ever intended to grow crops; archaeologists have found no horticultural tools or plant remains.[58] It is difficult to imagine that the people of the Middle Missouri Tradition would have believed that corn horticulture could succeed at that location.

The site may not represent an attempt to establish a permanent self-contained village at all, merely a camp that could be reoccupied seasonally. Archaeologists found pottery sherds similar to those of the Middle Missouri Tradition at the site, but recovered no tools associated with skin preparation, butchering, or woodworking. The effort expended in construction nonetheless suggests that the builders intended it to be more than a transitory camp. The site was probably occupied for no more than a few months, perhaps during a single spring. The Cluny people clearly relied on the bison for food. Evidently they killed the bison in the field and brought the choicest and most portable cuts to the village. They used these cuts very thoroughly, even shattering bones to extract marrow.[59]

Cluny was not unique. It is only the most northwesterly of several mud house sites on the northwestern plains that are known in the documentary and archaeological record. In 1800 Peter Fidler saw "at a very great place for buffalo crossing" on the South Saskatchewan River that there were "3 *Mud houses* on this side amongst a few poplars, they are of a circular form about 9 feet diameter & 4 1/2 high, they appear to be nearly 20 years old, they are said to have been built by a small war party from the Mis sis soury river, who live in these kind of habitations."[60] In 1802 Charles Le Raye saw what he understood to be Crow winter camps on the Bighorn River. He described them as "sunk three feet below the surface of the ground, but otherwise are built nearly similar to those of the Gros Ventres [Hidatsas]."[61] In October 1805, in a coulee near the confluence of the Yellowstone and Missouri Rivers, François-Antoine Larocque discovered "a lodge made in the form of those of the Mandans & Big Belly's [Hidatsas] (I suppose made by them) surrounded by a small Fort. The Lodge appears to have been made 3 or 4 years ago but was inhabitted last winter. Outside of the fort was a kind of stable in which the[y] kept their horses. There was plenty of Buffalo heads in the fort, some of them painted Red."[62]

The One Gun phase probably was not a transition from sedentary horticulturalism to nomadic hunting so much as a transition from pedestrianism to a nomadic equestrianism particularly suited to the environment of the northwestern plains. Inspired by their roots on the middle Missouri, the One Gun communities may have tried to develop semipermanent, seasonally occupied villages established at strategic locations in valleys where bison were common. Perhaps they assumed that by allowing hunters to transport bison to the villages from greater distances horses would enable them to establish a semisedentary bison-hunting way of life. In fact, as happened elsewhere on the Great Plains, equestrianism worked to the advantage of nomadic rather than sedentary groups, and the innovation of the One Gun peoples did not long survive the equestrian transition. Although Cluny was occupied only briefly, artifacts found there suggest that the villagers' connections to the Missouri villages had weakened. Knife River flint, available near the Hidatsa villages, is rare or absent. Those who built the villages did not carry their tools from the middle Missouri but manufactured them from local materials or, rarely, from obsidian that they gathered or traded from its source at the headwaters of the Yellowstone River.[63]

Throughout the pedestrian era diverse human communities migrated onto the northwestern plains. The environment militated against the development of horticulture in the region but permitted distinctly different bison-hunting subsistence economies. Residents of the region struggled and cooperated among themselves and participated in trade networks that extended far beyond the northwestern plains. At least one successful group of hunters lived in a symbiotic relationship with horticulturalists on the middle Missouri. Even if it is impossible to reconstruct the dynamic relationships that must have existed between human communities, we know that trade, warfare, and diplomacy are ancient.

The diversity that archaeologists have discovered is especially impressive, because archaeological evidence tends to obscure short-term developments and to mask the ethnic variability possible within a single archaeological culture. Although only a few archaeological cultures dominated the northwestern plains on the eve of the equestrian era, we know that many ethnic groups speaking dialects of Algonkian, Siouan, Numic, Salishan, Athapaskan, and Kutenian languages lived in the region at that time. Knowledge of the history of these various communities is essential to an understanding of the history of the northwestern plains during the equestrian era.

CHAPTER FOUR

Migrants from Every Direction

Communities of the Northwestern Plains to 1750

The Indians of the Plains are of various Tribes and of several languages which have no affinity with each other.

The Stone Indians are a large tribe of the Sieux Nation. . . . They have always been, and are, in strict alliance with the Nahathaways. . . . The Fall Indians, their former residence was on the Rapids of the Saskatchewan, about 100 miles above Cumberland House. . . . The Sussees are . . . brave and manly. [T]he three tribes of the Peeagan . . . all speak the same tongue, and their hunting grounds [are] contiguous to each other; these were formerly on the Bow River, but now [extend] southward to the Missisourie.

All these Plains, which are now the hunting grounds of the above Indians, were formerly in full possession of the Kootanaes, northward; the next the Saleesh and their allies, and the most southern, the Snake Indians and their tribes.

—DAVID THOMPSON

Documentary sources are useful for interpreting the history of the northwestern plains from the eve of the equestrian era. Like the archaeological evidence, however, the documentary material is fragmentary, discontinuous, and often apparently contradictory. Only two archaeological phases, Old Women's and One Gun, dominated the northwestern plains at the dawn of the equestrian era (see figure 4.1). The Mortlach and Selkirk phases reached their maximum westward expansion on the northern

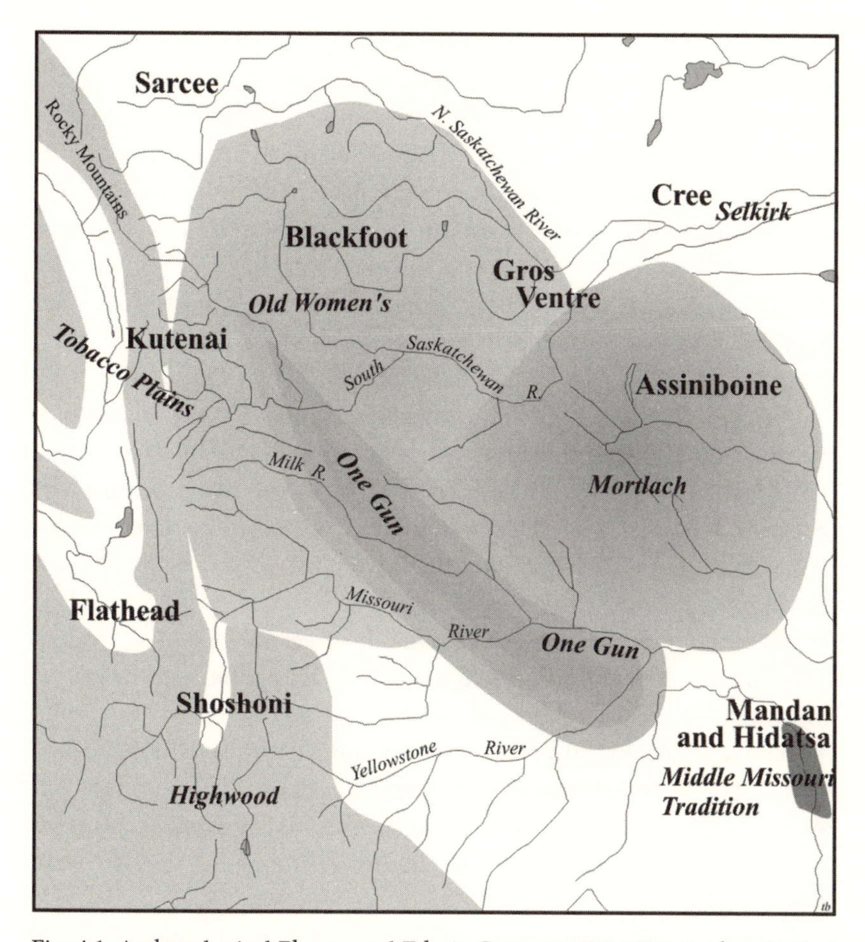

Fig. 4.1. Archaeological Phases and Ethnic Groups, 1700–1750. Archaeological phases are indicated by bold italics and historic ethnic groups by large bold letters. Adapted from Walde, Meyer, and Unfreed, "Late Period"; and Magne, "Distributions," 224.

plains near the end of the pedestrian era, but even then they were limited to the periphery of the northwestern plains. The documentary evidence reveals much more of the linguistic and cultural diversity of the region than the archaeological evidence does.

Scholars have long attempted to harmonize the documentary and archaeological evidence; but their attempts have produced some hotly contested debates, in part because archaeological cultures are not equivalent to anthropological cultures. Although we know that human societies can diverge, split, fuse, and amalgamate, the archaeological record often gives little evidence of these processes. Archaeologists cannot distinguish between the material culture of the Mandans and Hidatsas, for instance, even though they were distinctly different peoples. Conversely, groups belonging to a single ethnic group might have had very different material cultures. Individuals and groups of one community could be integrated into and contribute to the development of "foreign" groups. Virtually all human societies have complex roots in various societies. For example, the contemporary Blackfoot people almost certainly have ancestors among the Kutenais, Flatheads, Shoshonis, Crows, Crees, Nez Perces, and Sarcees, just as most of these groups have some Blackfoot ancestors. We must recognize and acknowledge this complexity in our ongoing efforts to reconcile the archaeological, linguistic, and documentary evidence. It is clear that new communities arrived on the northwestern plains continually and that all of them were influenced by their wide-ranging interactions with other groups.

Several Algonkian groups, including the Blackfoot, Gros Ventre, and Cree bands, occupied the northwestern plains during the early equestrian era. Most linguists argue that all of them originated in the proto-Algonkian homeland in what is now southern Ontario (although some argue that this homeland encompassed the entire region southwest and west of Hudson Bay) around 4000 B.C. The Algonkian speakers migrated to the northwestern plains from the northeast or east, but they did so at different times.

There is a near consensus among scholars that the historic Blackfoot people developed from the Old Women's phase. The archaeological, linguistic, oral, and written evidence supports this conclusion. While central and eastern Algonkian languages (like Cree, Ojibwa, and Micmac) are very similar, the plains Algonkian languages (Blackfoot, Cheyenne, and Arapaho/Gros Ventre) are very different from each other and from the central Algonkian languages.[1] This suggests that the Blackfoot people were separated from other Algonkians for many centuries. The distribution of the Old Women's phase at the end of the pedestrian era matches the distribution of the Blackfoot bands when the Europeans first encountered them, suggesting that their ancestors probably left most of the artifacts that we now categorize as Old Women's ware. The Blackfoot bands probably have ancient roots on the northwestern plains. Other Algonkian bands, the Gros Ventres and Crees, migrated to the region more recently. Unlike the Blackfoot bands, some of their bands had strong connections with other communities far removed from the area.

The history of the Gros Ventres is obscured in both the archaeological and early documentary record, probably because this relatively small linguistic group seems always to have had important interethnic connections with several other groups. Euroamerican traders are likely to have met Gros Ventre people before the 1770s, when documentary records first specifically identified them. In the 1770s the Gros Ventres occupied the region near the forks of the Saskatchewan River. The Ojibwa and Cree name for the Gros Ventres, which the traders translated as "Fall Indians," "Waterfall Indians," or "Rapids Indians," alluded to the strong current found in the Saskatchewan River system near the forks. Archaeologists have uncovered no archaeological remains on the northwestern plains that could be unique to the Gros Ventres.

The Gros Ventre and Blackfoot bands were closely associated during the equestrian era. If their friendships developed during the pedestrian era, the ancestors of the Gros Ventres may have adopted and contributed to the Old Women's phase. Whether or not the Blackfoot and Gros Ventre material cultures merged by the beginning of the equestrian era, their affiliation was so close that observers and historians have frequently categorized the Gros Ventres as part of the "Blackfoot Confederacy." While that term is misleading, the close relationship between the Blackfoot and Gros Ventre bands by the 1770s is crucial for understanding Gros Ventre history. Observers and historians have placed much less emphasis on the relationship between the Gros Ventres and Arapahos, but this relationship is also important for understanding Gros Ventre history.

Despite their affiliations, the Gros Ventre and Blackfoot bands had very different languages and very different pasts. Gros Ventre was mutually unintelligible with any other Algonkian language except Arapaho. Arapaho and Gros Ventre are dialects of a single language. According to Arapaho oral tradition recorded early in the twentieth century, the Gros Ventres were merely the northernmost of the five closely related Arapaho groups.[2] The Gros Ventres and Arapahos certainly have common roots.

While it is impossible to reconstruct the history of the Gros Ventres and Arapahos conclusively since their separation, linguistic and archaeological evidence, combined with oral traditions, makes it reasonable to postulate that both groups developed from the Duck Bay phase, which appears to have emerged around A.D. 1250 in the interlake region of present-day Manitoba from the Blackduck culture.[3] Linguistic evidence and the traditions of each group suggest the paths they took. The ancestors of the Gros Ventres became residents of the Saskatchewan River forks region, probably by 1550.[4] Their linguistic kin migrated toward the southwest.

The Arapahos may have settled in horticultural villages for a time before continuing their migration.[5] Perhaps it was on the northeastern plains that they first encountered the Cheyennes and continued their southwestward migrations in company with Cheyenne bands. The association between Arapaho and Cheyenne bands is very old. The Arapahos in the late nineteenth century professed to have no memory of a time when they were not closely affiliated with the Cheyennes.[6] Although Arapaho and Gros Ventre bands eventually formed affiliations with Cheyenne and Blackfoot bands, respectively, the Arapaho and Gros Ventre languages resemble the Eastern Algonkian languages (Ojibwa and Cree) more than they do the other Plains Algonkian languages (Blackfoot and Cheyenne). At some point the Arapahos divided again, with the Northern Arapahos eventually establishing themselves in the upper Platte River region.

Arapaho and Gros Ventre bands maintained contact with one another throughout their separation, although the significance of their contacts may have ebbed and flowed during the pedestrian and equestrian eras. Through their Arapaho kin, the Gros Ventres also had friendly relations with the Cheyennes during the equestrian era.[7] The Gros Ventre bands had friendly relations with other indigenous groups far outside the northwestern plains, which became tremendously important soon after Euroamericans arrived, when the military and diplomatic position of the Gros Ventres deteriorated. Despite their residence on the northwestern plains and their association with Blackfoot bands, the Gros Ventres had greater access to the southern Great Plains than any other group in the region.

Today archaeologists agree that the modern western Crees developed from the archaeological tradition known as Selkirk. This knowledge has forced historians and archaeologists to reexamine an earlier assumption that the Crees expanded westward into the Saskatchewan River basin and the western forests in response to factors related to the Euroamerican traders, such as game depletion.[8] The Selkirk peoples resided in the

Saskatchewan River Valley below its forks by the 1400s and were expanding westward in the years thereafter. The expansion of the Crees into the forests and parklands north of the Great Plains clearly predates their contact with Euroamericans, although the archaeological evidence does not rule out subsequent migrations of Crees into areas occupied by Selkirk bands before 1600. Cree individuals and bands continued to migrate westward from the Lake Winnipeg region after Euroamericans arrived. Unless western Cree bands were also producing non-Selkirk archaeological artifacts, however, there is no reason to question the long-held belief (based on documentary evidence and oral traditions) that Cree bands became permanent residents of the northern plains only during the late pedestrian or early equestrian era.[9] The Selkirk peoples had long-standing north-south trading relations that extended as far as the middle Missouri villages.

All Siouan speakers apparently originated in a proto-Siouan homeland toward the southeast, but several of them are important in the history of the northwestern plains. The history of various Siouan communities on the northern plains is certainly related to the Middle Missouri Tradition, Mortlach phase, and One Gun phase, but scholars have interpreted the archaeological and documentary evidence in several ways. Few doubt that the people who established the middle Missouri villages about nine centuries ago included ancestors of the Hidatsas. These villages exerted a significant influence on the northern plains in both the pedestrian and equestrian eras, primarily in their role as trading centers.

Less significant than their role as traders, but nonetheless important, was the direct role that the Hidatsas played in the history of the northern plains. Documents exist to place Hidatsa hunting communities on the northern plains in the late pedestrian era. In 1691 HBC employee Henry Kelsey traveled inland from York Factory to encourage the "Naywatame Poets" to travel to Hudson's Bay to trade. On the northeastern plains he met tents of Naywatame Poets; Dale Russell has argued that these were Hidatsas.[10]

The clear evidence that the Hidatsas were not restricted to the villages and that some of them traveled well onto the northeastern plains in the pedestrian era has led some to suggest that the Mortlach phase represents this Hidatsa presence. Mortlach, however, is probably much more complex than this. About 10,000 people lived in the Mandan and Hidatsa villages at the Knife River and Heart River in the late pedestrian era. Given that resources became depleted near the villages even after smallpox reduced their populations in the 1780s, it is safe to assume that they did so earlier, when those villages were larger.[11] Even in normal years Hidatsa hunting parties likely traveled considerable distances to hunt and were absent from the villages for several months at a time.[12] Some Hidatsa bands that depended wholly or largely on the plains hunt may have participated in a symbiotic relationship with the horticulturalists throughout the history of the villages.

Hidatsa hunters were unlikely to have been primarily responsible for leaving Mortlach artifacts. The ancestors of the Assiniboines, another Siouan group, probably hold this distinction. The Assiniboines have a very different background from the Hidatsas. Linguistically the Assiniboines are very closely related to the Nakotas, also known as the Yanktonais or Yankton Sioux. While scholars formerly believed that the Assiniboine/Nakota schism occurred in the seventeenth century, it is becoming increasingly evident that the division occurred considerably earlier. Documentary evidence shows that by 1700 Assiniboine bands occupied the parklands and forests of the lower North Saskatchewan River area,[13] where they had friendly relations with Cree bands. In 1754 Anthony Henday met "Eagle Indians" (an Assiniboine band that at this time "never had traded with any European or Canadian") on the margins of the northwestern plains.[14] The earliest Euroamerican observers identified the North Assiniboines, oriented toward the parklands, and the South Assiniboines, more oriented toward the grasslands. The two groups seem to have had markedly different cultures.[15]

The evidence of the distribution and the internal diversity among the Assiniboine bands by 1700 certainly suggests that the Assiniboine presence on the northern plains considerably predates that time. The congruence between the known positions and heterogeneity of the Assiniboine bands during the early equestrian era and the distribution and diversity of the Mortlach peoples has led some scholars to conclude that the Assiniboines developed out of the Mortlach phase.[16] This is significant, for it suggests that the ancestors of the Assiniboines participated in extensive trade with both Cree and Hidatsa communities for several centuries before Euroamericans arrived. The alternative explanation, that the Mortlach phase represents a significant Hidatsa presence on the northern plains, suggests that the Assiniboines rapidly invaded the northern plains upon the withdrawal of the Hidatsas as Euroamericans arrived.

The Hidatsas do appear to have been under considerable pressure when Euroamericans came to the area. In the late seventeenth century Henry Kelsey explained that Cree and Assiniboine bands were at odds with Naywatame Poets on the northeastern plains.[17] It is likely that these Cree and Assiniboine bands were protecting their privileged access to European goods on the Hudson Bay. Pressure from rival groups, armed from the north by the Hudson's Bay Company and from the east by the French, limited Hidatsa access to the northeastern plains in the late seventeenth century and early eighteenth century. This may partly explain the growing presence of middle Missouri villagers on the northwestern plains at the same time.

The Crows were a Siouan-speaking community of hunter gatherers that developed from the Hidatsas. Their emergence on the northwestern plains, which dates from the late seventeenth or early eighteenth century, may be tied to the arrival of the gun on the northeastern plains and the horse on the northwestern plains. Both the Crows and the Hidatsas were present on the northwestern plains in the early equestrian era, the Crows as residents and the Hidatsas as warriors and raiders. They may have left archaeologically identical artifacts, including mud houses, at the same locations during the same period. The One Gun phase must certainly represent the Hidatsa / proto-Crow / Crow presence on the northwestern plains between 1675 and 1750.

Some of the earthlodge villages on the northwestern plains probably represent the Hidatsa response to resource depletion near the villages. Perhaps their inability to continue hunting on the northeastern plains forced them to travel upriver instead. Traveling only with essentials, villagers could resort to the mud houses seasonally to alleviate the demand on food stores at the villages and return to the villages when their labor was required. If they were fortunate, they could return with some meat. As their northeastern neighbors acquired guns, the Hidatsas may have withdrawn from the northeastern plains region and sent their hunting expeditions toward the west. When these migrant Hidatsas on the northwestern plains acquired horses in the early eighteenth century, they may have begun to form a separate identity that Westerners came eventually to know as the Crows.

The Crow ethnogenesis nearly coincided with the beginning of the equestrian era. Several of the most prominent students of Crow history have concluded that the emergence of the Crows on the northwestern plains occurred after 1650.[18] John Ewers argued as early as 1960 that the Crows emerged in the eighteenth century, after the arrival of the horse but before 1776.[19] Archaeologists Raymond Wood and Alan Downer came to the same conclusion using ethnohistorical evidence. The reports of the experienced trader Edwin Denig suggest that the division took place between 1750 and 1775. An estimate by James H. Bradley in the late nineteenth century gives the same approximate date.[20] Flathead informants told James Teit that the Flatheads, who would later develop close relations with the Crows, first encountered them about 1750.[21] When Matthew Cocking enumerated the enemies of the Crees in 1772, he mentioned "four Nations, Kanapick Athinneewock or Snake Indians: Wahtee or vault Indians [Hidatsas] Kuttunnayewuck [Kutenais] and Nah-puck Ushquanuck or flat Head Indians."[22] Cocking did not explicitly identify the Crows, which has led some scholars to assume that the Crows were or were included among the "Snake Indians."[23] It is more likely that Cocking's informants included the Crows among the "Vault Indians." Since Indians were still erecting earthlodges on the northwestern plains, Cocking's informants may not yet have learned (or did not explain) that some of these Vault Indians had formed a separate identity.

The Crows are also not mentioned in other HBC journals from the 1770s and 1780s. Strangely, however, the earliest reference to "Crow Indians" dates from 1716, when a "Mountain Indian" arrived with a "Cocauchee or Crow Indian" "slave" woman.[24] If this woman was a member of the Crow community in question, the emergence of the Crows must be dated earlier than has been suggested. It is a tantalizing possibility, but perhaps impossible to confirm.

If, as Arthur Ray has argued, the Mountain Indians (who were clearly aware of horticultural villages) were Hidatsas,[25] they certainly would have been aware of a Hidatsa-Crow schism before the Crees or Assiniboines were. They would also have been more likely to differentiate between themselves and the Crows. The fact that the Crow woman was a slave suggests that the Hidatsas and Crows may have been at war at the time. Perhaps the division was recent. While the relationship between the Crows and the Hidatsas was not always friendly in the equestrian era, there is little record of warfare. The emergence of the Crows, however, was probably linked to a rupture between the Hidatsas and their western kin. According to Charles McKenzie, the Crows in 1805 had a tradition that the Crow-Hidatsa separation occurred when two Hidatsa brothers were forced to flee the villages after murdering their own relations. They married Flathead women and developed a nomadic bison hunting lifestyle and a separate dialect.[26] There are other accounts, among both the Crows and the Hidatsas, but they all suggest that the schism was accompanied by hostilities. The reference to the Crows in the HBC documents undermines the specific references to 1750 as the date of the schism and supports Ewers's argument that the arrival of the horse was the catalyst in Crow ethnogenesis.

The Algonkians and Siouans on the northwestern plains originated from the northeast and southeast, respectively. Other groups entered the region from different directions. Linguists tell us that the proto-Salishan homeland was probably in the lower Fraser River Valley and Puget Sound area. Similarities among Salish languages suggest that proto-Salishan groups spread gradually outward from there, maintaining contacts with one another.[27] One of these groups, the Flatheads, had a significant presence in the region. They occupied the northwestern plains only seasonally during the equestrian era and may not have occupied the area year-round even in the pedestrian era, although they did have close ties during the equestrian era with Shoshoni and Crow bands. They also occupied areas west of the Rocky Mountains that were capable of supporting much larger horse populations. The Flatheads were the source of many of the horses that ultimately found their way, through trade or warfare, onto the northern plains. The history of the Flatheads during the pedestrian era is difficult to trace because archaeologists have not been able to tie them solidly to an archaeological phase.

The history of the Kutenais seems even more difficult to trace, moving quickly from certainty to informed conjecture to mere speculation. Documentary evidence, strengthened by Kutenai traditions, makes it clear that some Kutenai bands resided east of the Rocky Mountains permanently until the 1780s.[28] Although there are contradictory traditions, long-standing and persistent traditions among the Kutenais indicate that they were originally a plains people.[29] In one of the earliest ethnological reports on the Kutenais, Alexander Chamberlain noted that "one of their myths ascribes to them an origin from a hole in the ground east of the Rocky Mountains."[30] Unfortunately there is no archaeological evidence from the late pedestrian and early equestrian periods that might shed light on Kutenai presence on the northwestern plains. A very reasonable argument ties the Kutenais west of the continental divide with the Tobacco Plains archaeological phase. This phase is restricted to the west side of the Rocky Mountains, however. Perhaps artifacts left behind by the plains Kutenai bands are now categorized as Old Women's ware. Some have suggested that the Kutenais developed out of the Avonlea phase,[31] which would mean that the ancestors of the Kutenais had an important place in the history of the northwestern plains.

Questions related to the Kutenai language and its relationship to other languages may hold the key to an understanding of important aspects of human interaction in the late pedestrian era. The Kutenai language is categorized as an isolate with no proven derivation, although this is not meant to imply that the language developed in isolation from all others. It merely means that Kutenai is too dissimilar from any other languages to be classified with them. Even if the genetic roots of the Kutenai language are proven, the language will probably still be classified as an isolate. Research in Kutenai linguistics is insufficiently developed to draw any firm conclusions, but studies do suggest that the Kutenais may well have deep roots to the west and to the east. Some promising studies suggest that the Kutenai language is genetically related to the Salishan languages but assumed its uniqueness as a result of many years of interaction between the ancestors of the Kutenai and Blackfoot peoples.[32] The resolution of this mystery would be of inestimable value for an understanding of the history of the northwestern plains in the late pedestrian era.

Anthropologists believe that all Athapaskan peoples have roots in the proto-Athapaskan homeland in northwestern North America near the Gulf of Alaska either along the coast or in the interior, where the widest diversity of Athapaskan languages is found.[33] Linguistic and archaeological evidence also suggests that ecologically devastating volcanic eruptions in the Saint Elias Mountains between 1,175 and 1,390 years ago may have triggered a series of Athapaskan dispersals in the years thereafter.[34] We know that ancestors of the Apachean Athapaskans, including the Navajos, migrated south from the northern boreal forests. These ancestors of the Apacheans almost certainly passed through the northwestern plains during the pedestrian era, but it is not yet possible to identify the archaeological record they may have left,[35] making it difficult to assess their impact. The role of the Sarcees, another group that made the transition from life in the northern boreal forest to life on the northwestern plains, is easier to trace.

Between 1700 and 1750 a group of Athapaskans separated from their Athapaskan kin, the Beavers, near Lesser Slave Lake.[36] Once on the plains the Sarcees quickly became tied to Algonkian bands. Their earliest associations were with Cree bands, but ultimately they became tied to the Blackfoot bands. Despite their different languages and history, the Sarcees forged such a close relationship with the Blackfoot bands during the equestrian era that some observers have even categorized them with the Blackfoot and the Gros Ventres in the "Blackfoot Confederacy."

Although we cannot explain why bands residing in the boreal forest migrated so quickly to a dramatically different environment where they could have had few kinship connections or why residents of the plains accepted them so easily, the Sarcee example illustrates that it was possible for a community to move onto the northwestern plains with little or no resistance from groups already resident there. The Sarcees did earn a reputation as fierce warriors, and Sarcee oral traditions related to Diamond Jenness in the early twentieth century suggest that their arrival on the plains was not entirely peaceful.[37] Still, they found common ground with prior inhabitants quickly, if not immediately, after they arrived. Had they failed to do so, they probably would have been forced to flee the region.

The Sarcees did not and probably could not simply battle their way onto the northwestern plains, but the Shoshonis evidently did. Although the Shoshonis are often associated with the early equestrian era, when they apparently briefly dominated much of the region militarily, their presence on the northwestern plains probably began around 1500. The Shoshonis were apparently already a formidable military force on the northwestern plains during the pedestrian era, even if they occupied only the southwestern margins of the region. The archaeological evidence of the Highwood Complex (assumed to represent the Numic spread onto the northwestern plains) emerges in the upper Missouri basin after A.D. 1500. The Numic expansion, like that of the Mortlach people, appears to have been aggressive. Some have argued that it may have been made possible by a broad-based subsistence strategy and a sociopolitical pattern that emphasized strong military organization.[38] On the plains the richer environment encouraged the development of larger bands. By the time the Shoshonis acquired horses between 1690 and 1700, they were probably already established in the Missouri Basin.[39] The Shoshonis eventually developed peaceful relations with Siouan (Hidatsa / proto-Crow) hunting bands connected with the middle Missouri villages, perhaps because of mutually beneficial trading links.[40] The lack of archaeological evidence for the Highwood people north of the Missouri River, however, suggests that the Shoshoni presence there was slight during the pedestrian era.

While the Shoshoni bands on the northwestern plains forged cooperative relations with the Crows, they also had important connections with other Numic speakers outside the region. The entire history of the region in the early equestrian era cannot be understood without recognizing that by the 1690s the Numic bands were widespread throughout the Great Basin and beyond.

Aspects of the Numic spread have been the subject of scholarly debate, although there is general agreement that the Numic-speaking peoples were once limited to southwestern portions of the Great Basin but then expanded throughout the region before the arrival of Euroamericans. Most agree that the expansion of the Numas into the eastern Great Basin and toward the northwestern plains was relatively late.[41] Scholars disagree, however, about whether the Numic spread should be seen as the aggressive expansion of certain communities or the amalgamation of diverse communities into one linguistic whole. By around 1500 Numic-speaking peoples occupied the northern limits of the Great Basin and the margins of the northwestern plains.[42] At that time the upper Snake River region supported large bison herds.[43] Other Numic bands were expanding southeast toward the southern Rockies. By 1600 Numic-speaking peoples who exhibited little linguistic diversity occupied a large territory from the upper Yellowstone to the southern Rockies. The Shoshoni bands in the north naturally maintained contacts toward the south. During the early seventeenth century some of the southernmost Numic bands known as the Comanches acquired horses that the Spanish had recently brought to New Mexico. This put the Shoshonis in the perfect position to become the first bands on the northwestern plains to have horses.

The expansion of the Shoshonis onto the northwestern plains has also been the focus of debate. The historical documents refer to an aggressive people known as the "Snakes" or "gens du serpent" who expanded onto the northwestern plains in the early 1700s. Some scholars have suggested that the term "Snakes" did not refer a specific ethnic group. Others argue that the "Snakes" were Siouan speakers. The most widely accepted view is that allusions to the Snakes of the northwestern plains refer to Shoshoni bands.[44]

By 1700 the northwestern plains were home to a great diversity of human communities that included Algonkians, Siouans, Athapaskans, Numas, Salishans, and Kutenais. Linguistic evidence suggests that these groups originally entered the Great Plains from every direction. Some of them battled their way onto the northwestern plains, while others arrived with little resistance. The behavior of each of these groups during the equestrian era was influenced by its history and its relationships cultivated during the pedestrian era. Some recent migrants were heavily influenced by ties to communities far beyond the northwestern plains. All of these communities were prepared to defend their position there militarily even as they cultivated and maintained cooperative relations with other groups. Each had a unique history and connections and faced different challenges and opportunities when the horse arrived among the Numic bands of the Missouri basin near the turn of the eighteenth century.

CHAPTER FIVE

The Horse and Gun Revolution, 1700–1770

Napi went away far South and there saw horses coming out of the water of a big lake. He made a rope of buffalo-hair and ran up behind the horses and caught a little foal which he carried off on his back, the mother and a great number of other colts coming with the colt. He took some of the mares dung and rubbed himself all over with it and the mare at once took kindly to him so that he caught her and brought all those horses up [along with] her and gave them to these Indians who broke them and rode them.

—BLACKFOOT ORAL TRADITION RECORDED IN 1896

Researchers have long debated but never resolved the question of whether or not the arrival of the horse triggered fundamental changes in plains cultures.[1] Scholars have paid much less attention to the effects of the horse and gun on Great Plains warfare and diplomacy. Frank Secoy's pioneering study, published in 1953, remains the only general survey. In contrast to many of his contemporaries, Secoy downplayed culture as an important factor in plains Indian history. He maintained that necessity led all plains communities to adopt or attempt to adopt all successful military innovations. Their only alternatives were flight or annihilation.[2] More recent work echoes Secoy's findings. Historian Dan Flores has described the horse as "the chief catalyst of an ongoing remaking of the tribal map of western America, as native American groups moved onto the Plains and incessantly shifted their ranges and alliances in response to a world where accelerating change seemed almost the only constant."[3] The horse and the gun, introduced to opposite corners of the Great Plains in the seventeenth century, had become military necessities by 1800.[4]

The arrival of the horse and gun revolutionized patterns of human interaction on the northwestern plains between 1700 and 1770. The fortunes of every indigenous group on the Great Plains quickly became tied to access to horses and guns. In the southern part of the region, the Shoshonis, a minor presence before 1700, led an aggressive interethnic coalition of bands that capitalized on their privileged access to horses to dominate the region militarily by 1740. About that time the southern coalition experienced a reversal: their northern enemies, increasingly well supplied with horses and guns, took the offensive, forcing them to begin a retreat. This was a matter of military might. Groups with little access to the instruments of war inevitably melted away before those with secure access. The environmentally favored tract between the Red Deer River and Missouri River became the contested area between the coalitions.

In January 1598 Juan de Oñate and a party of about 500 Spanish soldiers and clergy, women and children, and over 7,000 horses, sheep, and cattle embarked from Mexico City to establish the first Spanish settlements in the valley of the upper Rio Grande.[5] The settlements of New Mexico, which soon came to be centered around the village of Santa Fe near the pueblos, lay hundreds of unsettled miles north of Mexico. Oñate, a member of a prominent Mexican silver-mining family, hoped in vain that his *new* Mexico would duplicate Mexico's incredible mineral riches. Because of the colony's orientation, isolation, and chronically hostile relations with the Indians of the Plains and Great Basin and the lack of a substantial market for furs in Mexico or Spain, the tiny population of New Mexico came to rely more heavily on stock raising and trade in Indian slaves than on trade in furs.[6] A small fur trade did develop in New Mexico, especially at the pueblos; but, fearing that any traded guns would merely be turned back on their empire, the Spanish prohibited the sale of guns to Indians.[7]

By raiding New Mexico and trading with settlers, neighboring Indians soon got their first European goods. The pueblo dwellers, Apaches, and Utes acquired horses in the early seventeenth century. The Spanish probably trained Pueblo and Ute slaves in horsemanship early in the history of New Mexico, and it is likely that escapees brought horse-handling expertise to their kin. A hundred and fifty years later horse ownership had spread throughout the Great Plains.[8]

Indian groups usually got their first horses in friendly trade, since that allowed the transmission of both the horses and the requisite handling expertise. Those who acquired horses tried to prevent enemies from doing the same and were also reluctant to trade horses to friendly neighbors unless they were satisfied with their own supply. Once possessed of horses and horsemanship, indigenous communities attempted to augment their herds through breeding, trading, and raiding. This initial spread of the horse through peaceful trade produced a spoke-like pattern of diffusion rather than waves: the horse spread more quickly among peaceful and related groups northward from New Mexico and slowly in the crucible of the plains (see figure 5.1).[9] The Apaches were the first to use horse ownership to dominate their neighbors. Located near New Mexico in 1600, Apache groups occupied large areas of the western Great Plains a century later.[10] Their dominance was short-lived, however. In the successful Pueblo Revolt of 1680 Indians captured many Spanish horses. This spurred the spread of horses among indigenous groups.[11]

Between 1690 and 1700 Numic speakers of the Great Basin acquired their first horses, probably from their linguistic cousins, the Utes.[12] They quickly adjusted to equestrian life and began a remarkable expansion.[13] Some of them, whom the Utes would call the "Komanticas" (Comanches), burst onto the southern plains, moving toward the source of the horses and the bison herds of the region.[14] The nomadic Comanches, apparently unheard of in New Mexico before the eighteenth century, displaced semi-sedentary Apache groups and dominated the plains of eastern New Mexico by the middle of the eighteenth century.[15] The Comanches and others continually traded and raided for Spanish goods and horses at Santa Fe and at Taos after 1759.[16]

By the 1720s some of the Comanches' Shoshoni kin were already on the northwestern plains. Although their expansion into the region apparently began when they were pedestrian hunters, it probably occurred primarily after they acquired horses. Indians typically used horses domestically before they used them militarily; but even when they used their horses only domestically the Shoshonis must have enjoyed important military advantages over their pedestrian rivals. Horses permitted the Shoshonis to form large bands, especially in the summer, so they could

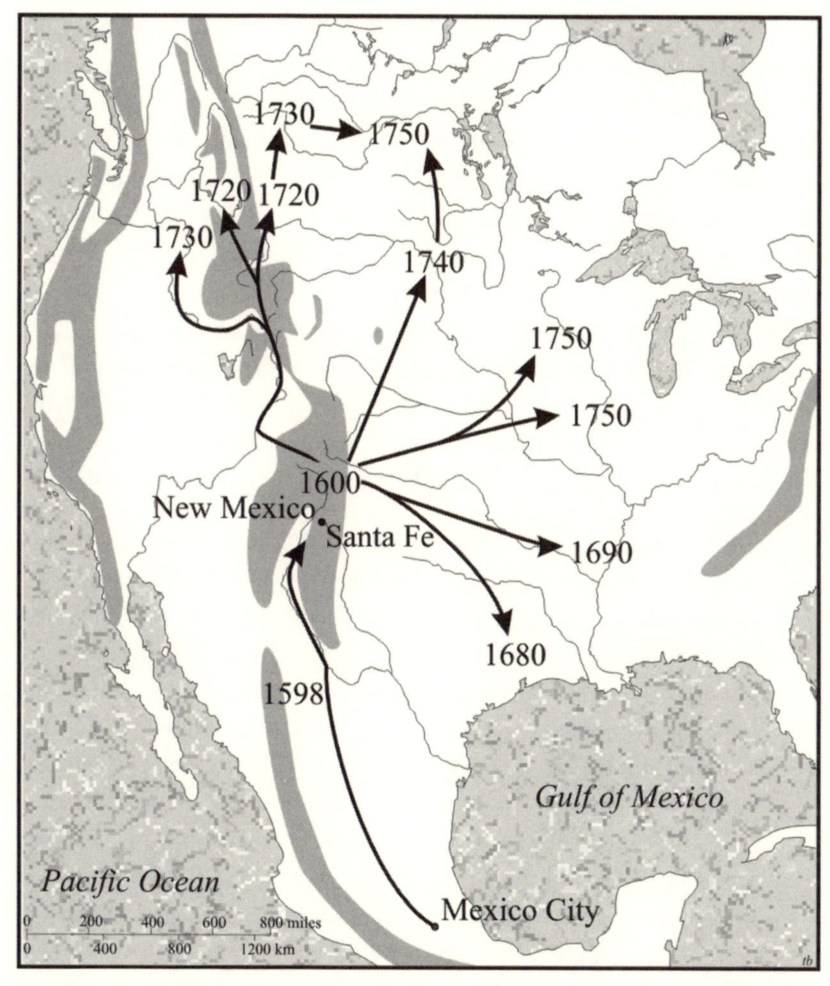

Fig. 5.1. The Spread of Horses on the Great Plains, 1600–1750. Adapted from Secoy, *Changing Military Patterns*; and Doige, "Warfare and Alliance Patterns."

muster larger war parties than their rivals. The Shoshoni bows and arrows were also noticeably more effective than those of their neighbors, in part because of their access to superior wood and lithic materials.[17]

The only historical record of "Snake"-Blackfoot warfare of the early equestrian era, and the most detailed historical account of plains warfare during this period, is contained in a fascinating although occasionally internally contradictory account preserved by David Thompson. Young

Man (Saukamappee) recounted his story in 1787, many years after the events occurred. He was born among the Crees of the Lower Saskatchewan in the early 1700s but became an influential Piegan leader as an adult. Young Man's story explains that the southern bands, without using horses in battle, forced the Blackfoot bands to flee from the South Saskatchewan River basin toward the Eagle Hills by about the 1730s, permitting the Shoshonis and the Crows or Crow-Hidatsas to dominate the northwestern plains.[18]

In their distress Blackfoot bands turned to their Cree and Assiniboine neighbors. By that time, when Young Man was a boy of about sixteen, the Blackfoot bands already knew about European goods, but possessed very few of them, if any. Even Young Man's Cree band had only a few guns and did not take them along when they joined the Piegans in battle:

> There were a few guns amongst us, but very little ammunition, and they were left to hunt for the families; Our weapons was a Lance, mostly pointed with iron, some few of stone, A Bow and a quiver of Arrows; the Bows were of Larch, the length came to the chin; the quiver had about fifty arrows, of which ten had iron points, the others were headed with stone. My [Young Man's] father carried his knife on his breast and his axe in his belt. Such was my father's weapons, and those with him had much the same weapons. I had a Bow and Arrows and a knife, of which I was very proud. We came to the Peeagan and their allies. They were camped in the plains on the left bank of the River (the north side) and were a great many. We were feasted, a great War Tent was made, and a few days passed in speeches, feasting and dances. A war chief was elected by the chiefs, and we got ready to march. Our spies had been out and had seen a large camp of the Snake Indians on the Plains of the Eagle Hill, and we had to cross the River in canoes, and on rafts, which we carefully secured for our retreat. When we had crossed and numbered our men, we were about 350 warriors . . . they had their scouts out, and came to meet us. Both parties made a great show of their numbers, and I thought that they were more numerous than ourselves.
>
> After some singing and dancing, they sat down on the ground, and placed their large shields before them, which covered them: We did the same, but our shields were not so many, and some of our

> shields had to shelter two men. Theirs were all placed touching each other; their Bows were not so long as ours, but of better wood, and the back covered with the sinews of the Bisons which made them very elastic, and their arrows went a long way and whizzed about us as balls do from guns. They were all headed with a sharp smooth, black stone (flint) which broke when it struck anything. Our iron headed arrows did not go through their shields but stuck in them; On both sides several were wounded, but none lay on the ground; and night put an end to the battle, without a scalp being taken on either side, and in those days such was the result, unless one party was more numerous than the other. The great mischief of war then, was as now, by attacking and destroying small camps of ten to thirty tents, which are obliged to separate for hunting.[19]

If the Shoshonis had horses at the time, they did not use them in battle. Soon afterward, however, thanks to their horses, the Shoshonis were the dominant military force on the northwestern plains. The kin connections between Comanches and Shoshonis, the lower incidence of warfare than in the plains, and the existence of the Shoshoni trade rendezvous in the upper Green River basin made the Great Basin a channel for horse transmission. By the 1730s Shoshoni bands had enough horses to begin trading them with Crow, Nez Perce, and Flathead bands. Before long the Kutenais also had horses.

Soon the Shoshonis began using horses in battle, leaving all their enemies reeling. On the northwestern plains the Shoshonis drove the pedestrian Blackfoot bands toward the North Saskatchewan River. Young Man's description hints at the effect that the sight of horse-mounted warriors had on the enemies of the Shoshonis:

> the Snake Indians and their allies had Misstutim (Big Dogs, that is Horses) on which they rode, swift as the Deer, on which they dashed at the Peeagans, and with their stone Pukamoggan knocked them on the head, and they had thus lost several of their best men. This news we did not well comprehend and it alarmed us, for we had no idea of Horses and could not make out what they were.[20]

Indian rock art preserves a record of the equestrian revolution. Figure 5.2 shows a petroglyph near the Milk River that commemorates a pedestrian

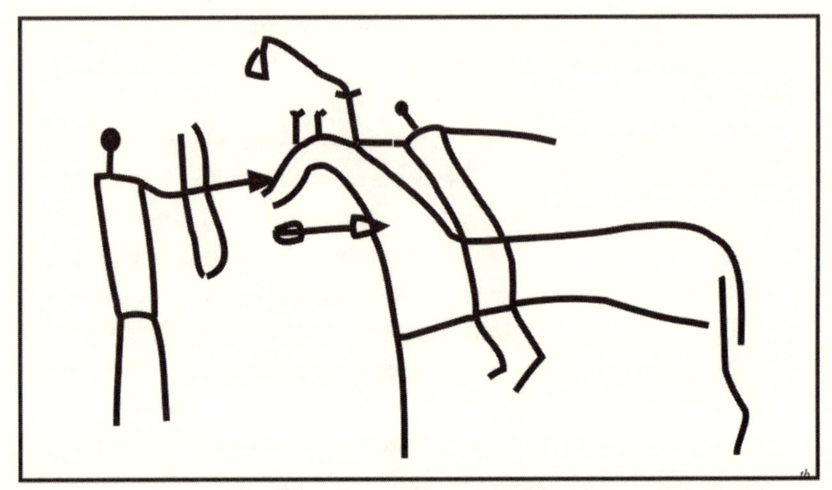

Fig. 5.2. Petroglyph of a Pedestrian and an Equestrian Warrior in Combat. A lightly armed pedestrian warrior meets an equestrian warrior in this petroglyph from Writing-on-Stone near the Sweetgrass Hills. The glyph evidently commemorates an instance when the pedestrian warrior was victorious over the equestrian warrior. Redrawn from Keyser, "Writing-on-Stone," 68.

warrior's success against a mounted opponent. Although such successes must have been exceptional, the glyph is a useful reminder that horse owners were not invincible in battle.

The Shoshoni and Crow bands simply did not have a large enough population to occupy or even patrol the territory they dominated. Their rivals continued to hunt in a broad contested zone, although always at the risk of Shoshoni attack, especially during the summers, when small pedestrian hunting parties were no match for the large, mobile mounted war parties. The dramatic military advantage that horses gave the Shoshonis encouraged their aggression. In 1742 Cheyenne informants told Louis-Joseph Gaultier de La Vérendrye that the Shoshonis

> do not content themselves in a campaign with destroying a village, according to the customs of all the savages; they keep up the war from spring to autumn. They are very numerous, and woe to those who cross their path! They are not friendly with any tribe. It is said that in 1741 they had entirely ruined seventeen villages, killed all the men and the old women, made slaves of the young women and sold them on the coast for horses and merchandise.[21]

Contrary to the Cheyennes' (or La Vérendrye's) assertion, the relations between the Crows and Shoshonis seem to have been generally peaceful during this era. The acquisition of horses appears to have led Crow bands to gravitate toward the excellent horse pasture of the Yellowstone basin. This may well have caused friction with some Shoshoni bands, but they seem to have continued to value Crow trade contacts with the Missouri villagers. With the addition of the horse, both as a mode of transportation and as an exchange item, trade relations between the Crows and the Missouri villagers became more important.[22] The Shoshonis maintained their access to the goods of the Missouri villages through Crow intermediaries.[23]

The arrival of the horse and increased trade contact between the Crows and Hidatsas, perhaps in concert with internal factionalism or population pressure among the Hidatsas, evidently encouraged new waves of Hidatsa migration to the northwestern plains. This was the period when migrants from the middle Missouri villages built and briefly occupied mud houses in the upper Missouri River basin and the South Saskatchewan River basin. The evidence that at least some of these people owned horses and a few metal tools suggests that the villages were settled, but soon abandoned, during the very early equestrian era.[24] The Crows or Hidatsas who lived in the villages probably found that a semisedentary lifestyle did not suit an equestrian people overwhelmingly dependent on the bison. Nevertheless, the fact that they settled as far north as the Bow River indicates that the bands of the southern coalition felt secure in that area in the early pedestrian era.

During the early eighteenth century the equestrian Shoshoni and Crow bands enjoyed such military advantages over their enemies that they dominated the northwestern plains. Flathead and Kutenai bands friendly with the Shoshonis also occupied the western margins of the plains.[25] Blackfoot and Gros Ventre bands retreated to the North Saskatchewan River region, although their hunting parties made forays farther south. Toward the middle of the eighteenth century, however, the fortunes of the Shoshonis turned. They could not prevent the spread of horses to the Blackfoot and Gros Ventre bands. Blackfoot groups, probably first the Piegans, likely acquired their first horses from Flathead or Kutenai bands in the 1730s.[26] The Piegans, in turn, must have shared their horses with their Blood and Siksika kin soon thereafter. Horses first arrived at the

Mandan/Hidatsa villages around 1740, perhaps in the company of Crow or Cheyenne traders. Various Cree and Assiniboine bands acquired their first horses from this trading center and from the Blackfoot bands. By the 1750s Cree and Assiniboine bands on the northeastern plains were probably among the last groups on the Great Plains to acquire horses.

The tide turned against the Shoshoni and Crow bands not only because the Blackfoot bands acquired horses, but also because they were simultaneously getting European weaponry. Blackfoot horse herds remained smaller than Shoshoni and Crow herds, but access to metal and to European weaponry allowed the Blackfoot bands to halt the Shoshoni expansion. The military importance of metal and guns induced the Blackfoot bands in the eighteenth century to cultivate friendly relations with Cree and Assiniboine suppliers. Cree and Assiniboine bands that visited Euroamerican traders forged symbiotic relationships with Blackfoot and Gros Ventre bands on the northwestern plains, providing them with European goods in exchange for horses.

After the HBC established a seasonal trading settlement at the mouth of the Nelson River in the 1670s and York Factory at the same place in 1684, European goods began filtering inland via indigenous trade networks. The British and French traders hoped to find precious metals in America's interior; but unlike the Spanish, they came primarily to trade the furs that the Indians supplied in exchange for a wide variety of European goods, including weaponry (see figure 5.3). Peaceful trading relations immediately developed between Euroamericans and Indians on Hudson's Bay.

Indians of the northwestern plains may have obtained some European goods before 1720, but they may not have acquired them in significant quantity until the 1730s. For the first few years after the establishment of trading posts on the shores of the Hudson Bay indigenous groups near the bay probably absorbed most of the supply of European goods, since many of these goods were consumed (tobacco, liquor, and ammunition), worn out (kettles), or quickly broken (guns). Until the 1730s, when traders from New France traveled into the region, the interior trade was handled solely by indigenous traders who had limited means or incentive to convey goods beyond their necessities. There certainly was no opportunity or incentive for individuals or bands to specialize as traders. Instead Cree and Assiniboine hunters acquired European goods for their own use

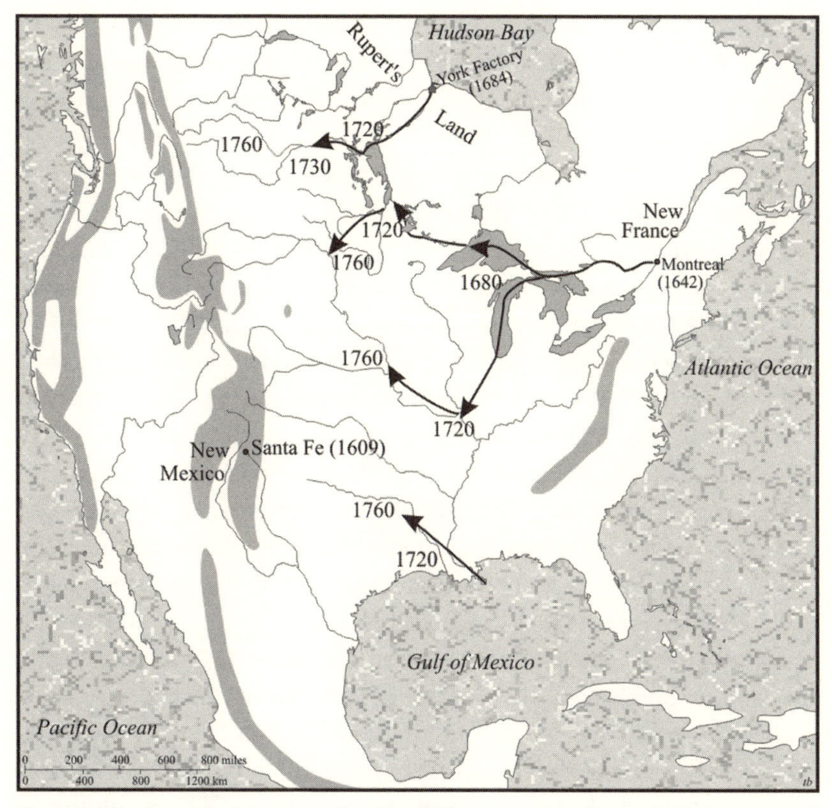

Fig. 5.3. The Spread of Guns on the Great Plains to 1760. Plains Indians acquired their guns either from the French traveling from New France and Louisiana or from the Hudson's Bay Company trading from Hudson Bay. Spanish settlements in New Mexico were an insignificant source of guns. Adapted from Secoy, *Changing Military Patterns.*

at York Factory in the summer and, shortly before returning to York Factory the following spring, traded these used goods with groups who did not make the trip to the bay. This method of trade limited the distribution of European goods, particularly goods like tobacco, liquor, powder, and shot, which were either exhausted or in short supply by spring.[27]

The uncertainty of supply also slowed the diffusion of goods before 1717. Unlike the supply of horses, which could grow naturally, European goods required constant replenishment from their source. The influx of European goods did generally become larger and more certain over the years, but trading Indians soon learned that this flow could be, and was, disrupted. Unless they were quite sure that they could replenish their stock of European goods at York Factory, Indian traders would want to retain a part of their own supply. This was an important consideration in the early years.[28]

The Indians of the northwestern plains could not have known about the geopolitical forces that interrupted the flow of European goods, but they were very much affected by them. The HBC established York Factory just as a period of relative peace between the English and French crowns was ending. The Glorious Revolution of 1688 ushered in seventy-five years of war and rivalry. France's Louis XIV would not allow the Protestant William and Mary to control the crown of England without a struggle. As part of the War of the League of Augsburg French forces captured York Factory in October 1694 and renamed it Fort Bourbon. The Treaty of Ryswick (1697) that ended that war called for the return of Hudson's Bay Company posts, but the French held Fort Bourbon until September 1714, when they finally relinquished it in accordance with the Treaty of Utrecht that ended the War of the Spanish Succession.

Fort Bourbon, operated by the French Compagnie du Nord, was poorly supplied with European goods. For example, the fort was not restocked at all between 1708 and 1713. This left the local Crees "in a bad way," and even the traders themselves without "enough food or powder to hunt game with guns."[29] Inland groups that traveled great distances to get to the bay stopped visiting in those years. Thirty canoes of "Mountain Indians" (probably connected to the Mandan/Hidatsa villages) visited York Factory after its restoration to the HBC in 1716 and told James Knight that they had not come to the bay for fifteen or sixteen years for fear of finding that the French would have no supplies for them.[30]

After the HBC regained control of York Factory, some inland Indians resumed trading there; but supply problems also plagued the HBC. In 1716 the annual supply ship failed to arrive at York Factory. These supply problems were not a mere inconvenience for the Indians who traveled to the bay. Their families who were left inland risked starvation and enemy attack. The Indian traders themselves risked traveling several weeks through a harsh environment to get to York Factory and back. If they arrived to find the post had not been supplied, they had to abandon their furs and return to their families empty-handed. Although these indigenous traders may have exaggerated the hardships they faced, their complaints to traders indicate the difficulties they endured. One leading inland Indian told James Knight in 1717 that because the supply ship had not arrived the previous year "he has Lost allmost all his Family and weeps very much & Said that A great Many Indians they have heard nothing of yet wch he fears is Lost Winter Comeing on upon them before they could gett into their own Country." The Mountain Indians explained that they had sealed a peace agreement with their enemies by giving them their guns, powder, shot, knives, and hatchets and, since they were now unable to replace these weapons, expected their neighbors to "kill 'em with their own weapons."[31]

Exaggerated or not, these reports explain why the inland Indians hesitated to travel to the bay until they were reasonably sure they could acquire supplies once they got there. Only about a third of the inland Indians who came to York Factory in 1716 came back in 1717.[32] These accounts also suggest why those who traveled to the bay hesitated to supply "strangers" with European manufactures, particularly before the 1720s. The flow of European goods to the northwestern plains probably was slight before 1720 and was limited largely to metal utensils and tools.

After the French gave up Fort Bourbon in 1714, they lost all access to the western interior via Hudson's Bay. They could compete with the British traders at the bay only by plying the waters of the Great Lakes. The French had occupied sites as far west as Kaministiquia (present-day Thunder Bay) in the late 1670s and early 1680s, but Louis XIV ordered them abandoned in the 1690s after the French captured HBC posts on Hudson Bay and fur prices plummeted because of a glut in the European market.[33] In 1714 the discovery that the surplus beaver furs in French warehouses had spoiled and the loss of Hudson Bay posts led French traders to return to those sites and beyond.[34] By the late 1720s the French were trading goods in York Factory's hinterland (see figure 5.4). French traders may not have made it to the plains in the 1720s, but their trading goods apparently did. A few Frenchmen who left the trade and were adopted by indigenous bands probably did meet Indians of the northwestern plains.[35] In the 1730s "Kis.ska.che.wan" (Saskatchewan) Crees began trading at York Factory.[36] These bands from the Saskatchewan River valley were probably neighbors of Blackfoot and Gros Ventre bands near the Eagle Hills. A party of "Bloody Indians (or Mithcoo Ethenue)" came to York Factory in 1733 in company with one of these bands.[37]

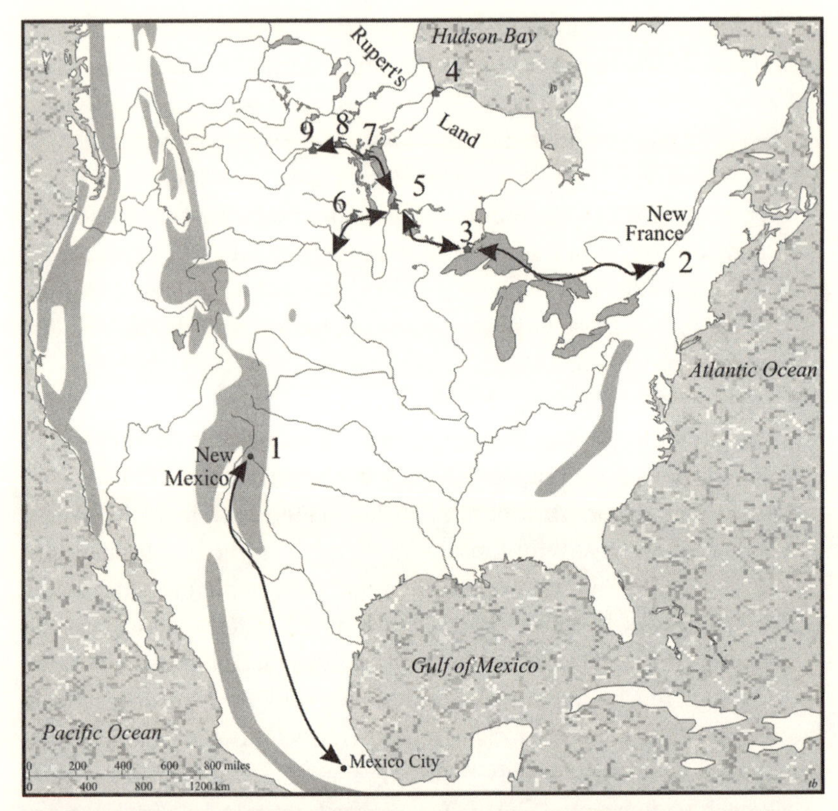

Fig. 5.4. Euroamerican Trading Centers and Transportation Routes to 1760. Beginning in the 1730s Indians of the northern plains were trading with French traders, who had expanded their operations westward from Montreal after 1714. The only significant trade with Spanish and Hudson's Bay Company traders was conducted by Indians who traveled directly to establishments in New Mexico and at York Factory. Legend: 1. Santa Fe (1609); 2. Montreal (1642); 3. Kaministiquia (1679); 4. York Factory/Fort Bourbon (1684); 5. Fort Maurepas (1734); 6. Fort la Reine (1738); 7. Fort Bourbon II (1741); 8. Fort Paskayoc (1750); 9. Fort à la Corne (1753).

On 4 June 1735 various Indians, including some "Kis.ska.che.wan" and Sturgeon Crees (probably from the lower Saskatchewan River region), confirmed a rumor brought to York Factory during the winter "of the french having fixd a Settlement on this Side ye Little Sea & that there was 10 Cannoes came to it this Summer, with trading goods & other necessaries."[38] This may be a reference to Fort Maurepas at the southern end of Lake Winnipeg, established in the summer of 1734 by Pierre Gaultier de

Varrennes et de la Vérendrye at the insistence of Cree and Assiniboine bands, although it could refer to an outpost established on the northern end of the lake by Gaultier's party or by other Frenchmen.[39] The company subsequently heard more news of the advance of the French traders in York Factory's hinterland. Sioux resistance slowed French expansion in the late 1730s, but by 1741 the French answered calls by certain Cree bands to establish a fort "at the entrance to the great English river" and built a new Fort Bourbon at the mouth of the Saskatchewan River.[40] During the War of the Austrian Succession (1744–48) a British blockade briefly interrupted French trade and expansion, but by 1753 French traders had established Fort à la Corne just downstream from the confluence of the North and South Saskatchewan Rivers at a major Cree gathering area known as Pehonan.[41]

The French dealt primarily in goods of high cost in relation to bulk. Their trading establishments were not substantial buildings; Anthony Henday described the "hogstye" that was Fort Paskoyac as a 26- by 12-foot log structure, with a birchbark roof.[42] Still, the French captured many of the furs otherwise destined for York Factory. The value of furs traded at York Factory peaked in the late 1720s.[43] French competition increased the flow of European goods to the northwestern plains. Cree and Assiniboine bands traded at these posts directly, and some of them conveyed goods west to the northwestern plains, where they traded them with Blackfoot, Gros Ventre, and other Cree and Assiniboine bands. The Blood and other "Earchethinue" (Strangers) bands that the HBC eagerly wanted to see at York Factory ceased visiting there.

It is probable, although impossible to prove, that Cree bands guided some Blackfoot bands to the French posts during these years. Blackfoot and Gros Ventre bands undoubtedly acquired valuable European products, including guns and ammunition, indirectly through their Cree and Assiniboine trading partners. A number of Cree leaders, among them White Bird (Wapinesiw) and Little Deer (Attickasish), emerged as important figures in the relationship with the Blackfoot groups.[44] It would be misleading to suggest a general Cree-Blackfoot alliance at that time. There were hostile encounters between Cree and Assiniboine bands and Archithinue bands in the 1750s and 1760s.[45] These hostilities were likely related to the earliest Cree and Assiniboine horse raids against the Blackfoot bands, which were destined to become the major irritant in Cree and Assiniboine relations with their neighbors in the years to come.

Relations between the Crees and Assiniboines and the Blackfoot groups were generally friendly in these years, and some Cree and Assiniboine bands were allied with Blackfoot bands. These alliances were strengthened by mutually beneficial trading relationships and mutual fear of the Shoshonis. White Bird and Little Deer had established cordial relations with the Blackfoot bands by the 1740s, following the pattern set earlier by Young Man's father. This relationship was not restricted to trade. Both White Bird and Little Deer joined Blackfoot forays against the Snakes.[46] Bands almost certainly intermarried, as they had in the 1730s. The cordial relations also allowed certain Cree bands to hunt in the upper North Saskatchewan River basin, including the Beaver Hills.[47]

The growing supply of European goods on the northwestern plains after 1730 dramatically affected the fortunes of communities. Between 1730 and 1750 the tide turned against the Shoshonis. After Young Man had spent a winter with his young wife, Blackfoot envoys came to ask the assistance of his Cree band. By that time the Crees of the lower Saskatchewan "had more guns and iron headed arrows than before," but the Blackfoot bands were hard-pressed by mounted Shoshoni warriors.[48] Young Man's description of the ensuing events reveals the tremendous impact that guns had on the Shoshonis:

> Only three of us went and I should not have gone, had not my wife's relations frequently intimated, that her father's medicine bag would be honoured by the scalp of a Snake Indian. When we came to our allies, the great War Tent [was made] with speeches, feasting and dances as before; and when the War Chief had viewed us all it was found between us and the Stone Indians we had ten guns and each of us about thirty balls, and the powder for the war, and we were considered the strength of the battle. After a few days march our scouts brought us word that the enemy was near in a large war party, but had no Horses with them, for at that time they had very few of them. When we came to meet each other, as usual, each displayed their numbers, weapons and shiel[d]s, in all which they were superior to us, except our guns which were not shown, but kept in their leathern cases, and if we had shown [them], they would have taken them for long clubs. For a long time they held us in suspense; a tall Chief was forming a strong party to make an attack

on our centre, and the others to enter into combat with those opposite to them; We prepared for the battle the best we could. Those of us who had guns stood in the front line, and each of us [had] two balls in his mouth, and a load of powder in his left hand to reload.

We noticed they had a great many short stone clubs for close combat, which is a dangerous weapon, and had they made a bold attack on us, we must have been defeated as they were more numerous and better armed than we were, for we could have fired our guns no more than twice; and were at a loss what to do on the wide plains, and each Chief encouraged his men to stand firm. Our eyes were all on the tall Chief and his motions, which appeared to be contrary to the advice of several old Chiefs, all this time were about the strong flight of an arrow from each other. At length the tall chief retired and they formed their long usual line by placing their shields on the ground to touch each other, the shield having a breadth of full three feet or more. We sat down opposite to them and most of us waited for the night to make a hasty retreat. The War Chief was close to us, anxious to see the effect of our guns. The lines were too far asunder to us to make a sure shot, and we requested him to close the line to about sixty yards, which was gradually done, and lying flat on the ground behind the shields, we watched our opportunity when they drew their bows to shoot at us, their bodies were then exposed and each of us, as opportunity offered, fired with deadly aim, and either killed, or severely wounded, every one we aimed at.

The War Chief was highly pleased, and the Snake Indians finding so many killed and wounded kept themselves behind their shields; the War Chief then desired we would spread ourselves by two's throughout the line, which we did, and our shots caused consternation and dismay along their whole line. The battle had begun about Noon, and the Sun was not yet half down, when we perceive some of them had crawled away from their shields, and were taking to flight. The War Chief seeing this went along the line and spoke to every Chief to keep his Men ready for a charge of the whole line of the enemy, of which he would give the signal; this was done by himself stepping in front with his Spear, and calling on them to follow him as he rushed to their line, and in an instant the whole of us followed him, the great part of the enemy took to flight, but some

> fought bravely and we lost more than ten killed and many wounded; Part of us pursued, and killed a few, but the chase had soon to be given over, for at the body of every Snake Indian killed, there were five of six of us trying to get his scalp, or part of his clothing, his weapons, or something as a trophy of battle.[49]

Young Man thought that the Shoshonis could have won despite their lack of guns but that this battle was the turning point in northwestern plains warfare in the eighteenth century because of the psychological impact of firearms:

> The terror of that battle and of our guns has prevented any more general battles, and our wars have since been carried by ambuscade and surprize, of small camps, in which we have greatly the advantage, from the Guns, arrow shods of iron, long knives, flat bayonets and axes from the Traders. While we have these weapons, the Snake Indians have none, but what few they sometimes take from one of our small camps which they have destroyed, and they have no Traders among them. We thus continued to advance through the fine plains to the Stag [Red Deer] River.[50]

Some of the drawings etched into the rock of the northwestern plains commemorate battles such as the one described by Young Man (see figure 5.5).

The ease with which one indigenous group could augment its horse herds by raiding another's herds did nothing to encourage friendships between horse-rich and horse-poor groups. Conversely, groups intent on maintaining constant and reliable access to European goods had to cultivate friendly relations with Europeans or intermediaries. For the Blackfoot bands this meant that cordial relations with Cree and Assiniboine suppliers were essential between 1730 and 1780. The Shoshonis had only the most tenuous trade connections with Euroamericans. Crow and Shoshoni bands acquired a few guns (almost inevitably without ammunition) from the Mandan/Hidatsa villages, or less frequently from New Mexico, but these were nearly irrelevant militarily. In 1770 the Shoshonis first used guns in battle with their northern enemies. White Bird, however, reported that they "cannot shote well yet" and had managed to kill only one Cree warrior at the expense of eighty of their own warriors.[51] Between the 1740s and the 1780s the Blackfoot bands and their allies

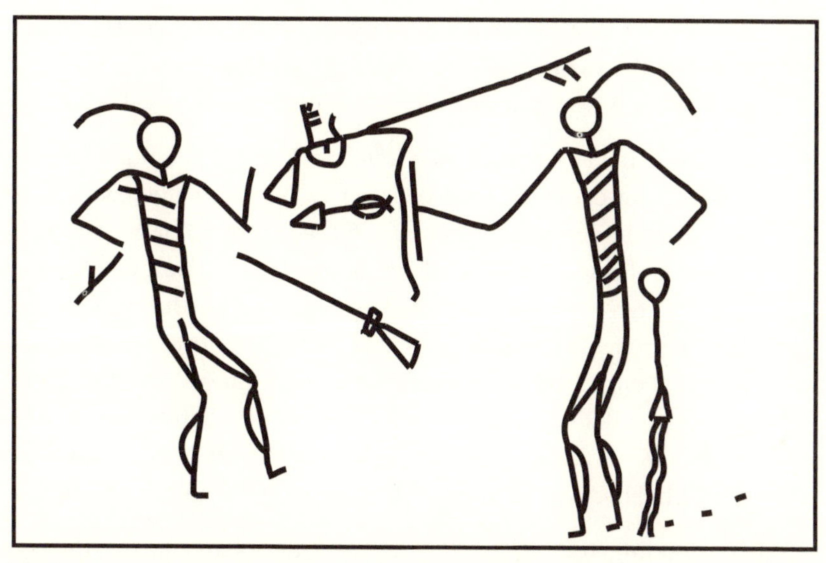

Fig. 5.5. Petroglyph of Combat in the Era of Euroamerican Weaponry. In this petroglyph the warrior on the right appears to have killed the enemy with his bow and arrow. He carries a coup stick in his left hand. Rock art often depicts warriors who were victorious despite facing better-armed foes. The original glyph is at Castle Butte near the Yellowstone River in Yellowstone County, Montana. Stuart Conner and Betty Lu Conner estimated that it dates from the 1740s. Redrawn from Conner and Conner, *Rock Art,* figure 14.

replaced the Shoshonis and their allies as the dominant military force on the northwestern plains.

On 14 October 1754 scouts from a large Archithinue camp near the Red Deer River approached a party of Keschachewan (Saskatchewan) and Pegogoma Crees led by Little Deer.[52] Among them were Anthony Henday, representative of "The Great Leader who lives down at the great waters," and Connawappa, a York Factory Cree leader.[53] The HBC had sent Henday "to ye Eachithnue, in order to bring to trade next summer."[54] From 14 to 17 October 1754, as the Archithinue camp grew from 200 to 322 tents (probably at least 2,000 people), the Archithinues listened as Henday tried to fulfill his mandate. The encampment's leader welcomed Henday and his party in a tent "large enough to contain fifty persons; where he received [them] seated on a . . . Buffalo skin, attended by 20 elderly men." Henday invited the young men to bring their furs to the bay "in return for Powder, Shot, Guns, Cloth, Beads, &c," but the leader "only said that it was far off, & they could not paddle."[55]

The following day Henday repeated his invitation, but the leader

> answered, it was far off, & they could not live without Buffalo flesh; and that they could not leave their horses &c: and many other obstacles, though all might be got over if they were acquainted with a Canoe, and could eat Fish, which they never do. The Chief further said they never wanted food, as they followed the Buffalo & killed them with the Bows & Arrows; and he was informed the Natives that frequented the Settlements, were oftentimes starved on their journey. . . . He made [Henday] a present of a handsome Bow & Arrows, & in return [Henday] gave him a part of each kinds of goods [he] had, as ordered by Mr. Isham's written instructions.[56]

After spending the winter with a small Cree band in the forests, Henday learned the routine established between various Archithinue bands and the Pegogoma and Saskatchewan Cree bands. In late March and April Cree and Assiniboine bands gathered birchbark and constructed canoes, probably along the North Saskatchewan River. At the end of April, after the river ice broke up, they embarked downriver. Between 12 and 21 May on the lower North Saskatchewan they met Archithinue (including Blood) bands with whom they traded the guns, kettles, hatchets, and knives they had gotten the previous summer for the wolf, beaver, and fox furs that the plains Indians had acquired during the winter. On 23 May Henday's hosts traded their most valuable furs at Fort à la Corne for the most valuable European goods then went on to York Factory with the bulkier furs, including most of the furs traded from the Archithinues, which they traded for bulkier goods.[57] Between 1754 and 1774 no fewer than fifty-six HBC men traveled inland with Cree and Assiniboine bands in the manner of Henday. Archithinue bands rarely visited trading posts themselves but met traders frequently and had secure access to European trade goods.

Even as contacts with HBC traders increased, the French presence in the region ended. When Henday departed York Factory for his inland voyage, events were already unfolding that would culminate in the elimination of French power in North America. On 4 July 1754 the twenty-two-year-old adjutant-general of the Virginia militia, Major George

Washington, and his men retreated from the upper Ohio River basin after surrendering Fort Necessity to the French. The Great War for the Empire had begun. Two years later the war spread to Europe, but the British war minister, William Pitt, decided that the British forces should fight this war with France by forcing a showdown in North America. That approach would affect many Indians, even those of the northwestern plains.

British strategy dovetailed well with the aims of the HBC. As the tide of the war turned against the French after 1757 and the British blockade of France became effective, French traders in the interior withdrew or did without their supplies from Montreal. For example, traders abandoned Fort à la Corne after 1757.[58] The war and subsequent uprising led by Pontiac cut off supply lines to the northwestern plains via Montreal, and the Indian inhabitants of the northwestern plains found themselves once again a thousand miles from the closest trading post. The value of furs traded at York Factory, which had declined significantly since the 1730s, rebounded in the late 1750s.[59] The continued profitability of the HBC and concerns about the navigability of the rivers gave HBC traders no incentive to establish inland posts.

With the French posts abandoned because of the Seven Years' War, the Indians of the northwestern plains turned back to the HBC as their sole supplier of European goods. Even the Blood Indians, who had ceased visiting York Factory after the French reached the Saskatchewan River, again visited occasionally between 1758 and 1763.[60] But Blackfoot visits ceased completely after 1763, when traders from Montreal returned to the Saskatchewan River. The Blackfoot bands were evidently quite satisfied with their arrangement with their Cree suppliers. They found the trip to the bay an avoidable annoyance.

An "Archithinue" man told Andrew Graham at York Factory that their Cree suppliers charged them fifty beaver or wolf skins for a gun. After being told that he could get a gun for fourteen Made Beaver (MB) at York Fort, he "generously told me [Graham] they never would come down, and that he himself never would come down again, as he did not like to sit in the canoe and be obliged to eat fish and fowl as he had done mostly coming down."[61] The long trip to the bay fit very poorly with the Blackfoot lifestyle. Since the Blackfoot bands did not travel by canoe themselves, they could visit the posts only with the guidance of their Cree suppliers, who must have expected compensation of some kind for their services. It is unlikely that they were saving themselves any expense by traveling to the bay. The Blackfoot bands seemed disinclined to take the risks associated with the trip. In 1763 Ferdinand Jacobs at York Factory reported that "the Bloody Indians, were Some of them Starved to Death Last year going Back which So intimidated them that I am afraid we Shall never have any more of them Come to Trade."[62]

That is the last recorded visit of the Bloods to York Factory. The reasons are clear. As early as 1766 HBC envoys began reporting the presence of Canadian "pedlars" on the Saskatchewan River. In 1768 a promising young Orcadian recruit, William Tomison, returned from an inland voyage reporting "a vast Number of Pedlars" in the interior.[63] In 1772 Matthew Cocking visited the "Archithinue" bands on the northwestern plains to invite them to York Factory; but he reported that "notwithstanding all I could say, promising they should be greatly rewarded; they were all unwilling alledging they were unwilling to leave their Families and feared a scarcity of Provision."[64]

The Blackfoot and the Gros Ventre bands may well have been fully satisfied with their arrangements with their Cree and Assiniboine suppliers in the 1760s. With Cree and Assiniboine aid the Blackfoot bands had turned back the southern bands by 1770 and felt secure in the Red Deer River basin. Further victories seemed assured. The HBC, however, could no longer afford to ignore the Canadian traders' encroachments. Reluctantly but deliberately the company established inland settlements, initiating an era of fierce Euroamerican competition for the trade of northwestern North America. As a result the Blackfoot and Gros Ventre bands established constant direct trading relationships with Euroamericans, which would have far-reaching consequences for all the Indians of the northwestern plains.

CHAPTER SIX

The Right Hand of Death, 1766–82

If one of our people offers you his left hand, give him your left hand, for the right hand is no mark of friendship. This hand wields the spear, draws the Bow and the trigger of the gun; it is the hand of death.

—YOUNG MAN
1787

This plaguey disorder [smallpox] by what I can hear was brought from the Snake Indians last Summer, by the Different Tribes that trades about this River, I can remember the time altho' it is but a few Years that they did not go to War above Once in three, but now they have got such great supplies of Ammunition &c. that they dont know what to do with it, they go every Year, if their had been None but Your Honors Settlements as usual, I dont think any thing of this kind would have fell amongst them.

—WILLIAM WALKER
1781

In the upper basin of the Rio Grande chronic warfare between Spanish settlers in New Mexico and the neighboring nomadic Indians reached a violent climax during the 1770s. In contrast the Euroamerican traders who were expanding their operations onto the margins of the northern plains at the same time established generally peaceful relations with indigenous groups. With an eye to their own profits, these traders also promoted peace among Indian groups on the northern plains. Ironically, as William Walker and other traders ruefully noted, the growing supply of guns and ammunition only encouraged more frequent and deadly warfare.

The history of the northern plains between 1766 and 1782 illustrates Alfred Crosby's pithy axiom that "whether the Europeans and Africans came to the native Americans in war or peace, they always brought death with them."[1] He was alluding to the effects of Old World diseases on American Indians, not the effects of European weaponry; it was smallpox transmitted to the Indians in New Mexico, not Spanish or British arms, that killed more people on the Great Plains between 1766 and 1782. The pandemic of 1781–82 afflicted every human community in the region except those of Euroamerican traders on the Saskatchewan River. It may be the most momentous event in the indigenous history of the Great Plains. As the epigraph shows, the perceptive William Walker understood that the first great smallpox pandemic on the plains was closely related to the horse and gun revolution that preceded it.[2]

The horse and gun revolution brought tremendous change to the northwestern plains before 1760 and thereafter, but indigenous communities also faced new challenges between 1766 and 1782. Changing conditions reinforced established patterns and foreshadowed dramatic realignments. By 1770 equestrianism had spread to Indian bands throughout the northwestern plains and beyond. Gun ownership had not; the growing disparity between gun-rich northern bands and gun-poor southern groups is one of the major themes of the period between 1770 and 1805. Existing patterns of warfare, trade, and diplomacy intensified. The coalition of horse-rich and gun-poor bands that dominated the northwestern plains between 1700 and 1750 was thrown on the defensive against the bands with secure access to Euroamerican weaponry. The large horse herds of the southern groups gave their owners little military advantage; they seemed only to invite enemy raiders. The supply of guns and ammunition that northern groups enjoyed was militarily decisive.

The northern bands waged the same kind of aggressive warfare against the southern coalition that the southern coalition had waged against them half a century before. Even in its ascendancy, however, the northern coalition began weakening. The generally cooperative relations among the northern bands endured, but the basis for these symbiotic relations eroded. Disagreements were more frequent, and friendships more difficult to renew. By 1782 divisions between Blackfoot, Sarcee, and Gros Ventre bands and the Cree and Assiniboine bands were becoming increasingly important.

Between 1766 and 1782 a significant Euroamerican presence reemerged on the northern plains, but Euroamerican expansion did not simply replicate conditions of the pre-1756 era. It unleashed new forces with important implications for some of the Indians of the northwestern plains. Equestrianism and the development of a significant market for plains provisions between 1766 and 1780 encouraged some western Cree bands to turn increasingly toward the resources of the plains. Cordial relations and kin connections between forest bands and plains bands eased their transition. Meanwhile, these very developments helped undermine friendly relations among these groups.

During this period Euroamericans moved westward to the margins of the northwestern plains. With the fall of Montreal in 1760, the HBC may have supposed that its monopoly of the trade of the northern plains was assured. In November 1762 France ceded Louisiana to Spain to prevent it from falling into English hands, but the Spanish could pose no threat to the company's interests. New France passed to the British. HBC officials had reason to believe that the Crown would safeguard the company's exclusive trading rights in Rupert's Land. In reality Montreal-based traders mounted a more serious challenge to the HBC after the fall of New France than they had during the French era.

Freed from French imperial restrictions and the limited capital available in New France and incorporated into the British Empire, traders in Montreal gained access to British manufactures and to the Brazilian tobacco that the northwestern Indians so preferred over the tobacco that the French had traded. For a few years before and after the fall of New France the HBC did enjoy a monopoly. The value of furs traded at York Factory grew steadily in the late 1750s and early 1760s before plunging again after 1763.[3] The effects of imperial struggle channeled furs and Indians toward York Factory between 1756 and 1763; but when the imperial struggle ended, furs began flowing back toward Montreal. HBC emissaries had little influence over Indian traders on the plains both before 1763 and afterward. When Canadian traders brought European goods to the Indians on the Saskatchewan River, few furs made their way to York Factory. Not only the Blackfoot traders were reluctant to visit the bay.

Prominent Cree trader White Bird, who began trading at York Factory in 1755, stopped coming in 1770. He preferred dealing with Canadians on the Saskatchewan rather than the HBC at the bay.[4]

The Canadians appeared around 1766 after the route from Montreal to the Saskatchewan River was pacified. In 1768 François Leblanc, a veteran of the French trade, built a post at "Nipowiwinihk" (Nipawin) not far downstream from the abandoned Fort à la Corne.[5] William Tomison reported "a vast Number of Pedlars" in the interior that year.[6] The Canadian posts, like the French posts established in the 1740s and 1750s, were located at important Cree rendezvous centers.[7] Unlike the French posts, however, these posts eventually forced the HBC to establish settlements in the interior.

At first the HBC responded to the renewed competition as it traditionally had. Each year it hired Cree and Assiniboine leaders to guide its traders to the interior. These emissaries invited Indians to the bay. This practice was counterproductive by the late 1760s, however. In 1768 Ferdinand Jacobs remarked that the HBC's envoys were unable to convince Indians to bypass the Canadian traders, but the envoys were in a perfect position to conduct a clandestine private trade with Indians in the interior to the company's detriment. Jacobs understood that the only solution for the HBC was to build trading establishments inland, where it could cultivate relationships with the Indian leaders like White Bird who directed a large part of the trade.[8] The lack of reliable men, however, prevented the HBC from carrying out the project until after 1772–73, when Matthew Cocking, one of the company's most talented servants, made an inland journey.

Isaac Batt was one of the more frequent HBC visitors to the northwestern plains in the 1750s and 1760s. He was inland most years between 1758 and 1773 and thus was familiar with the Canadian activities.[9] In May 1773 Batt was in England visiting with the London Committee of the HBC, which was alarmed by the decline in York Factory trade. He probably convinced the committee of the need for inland settlements, for on 26 May 1773, "upon mature Deliberation," the London Committee decided that "it would be to the Advantage of the Companys Trade to establish a Settlement Inland at or near Basquiau." It ordered Samuel Hearne and Matthew Cocking to establish the post and hire seventeen additional men at the Orkney Islands to supplement the company's personnel.[10]

Hearne and Cocking established Cumberland House in 1774 and renewed acquaintance with White Bird there. By then, however, rival Canadian traders were already competing with each other considerably farther upstream (see figure 6.1). Short of canoes and personnel, the HBC expanded its operations slowly at first.[11] The company depended on Cree men like Mameek Athinnee, who despite his insolent behavior was one of the few Indians willing to provide certain services to the company. The HBC also continued sending its envoys with bands farther into the interior. In the winter of 1776–77, for instance, the promising young recruits Scotsman Robert Longmoor and Orcadian Malcolm Ross spent the winter among plains Assiniboines at the Eagle Hills.[12] Still, throughout the interior the Canadian traders took the lead.

While a shortage of canoes and personnel crippled the HBC expansion, Canadian expansion was hindered by ruthless competition and their arrogant approach toward Indians. In the later 1770s rival Canadian traders reduced competition among themselves. The uncertainties caused by the United States War for Independence led to larger partnerships by 1776, but cutthroat competition still slowed expansion significantly before 1779.[13] In 1778 several traders pooled their resources to support a risky journey led by Peter Pond. When Pond returned from this fabulously successful expedition to the Athabasca, many Canadian traders recognized that they could not tap the fur riches of the Athabasca profitably unless they reduced competition and pooled their capital. This process gave birth to the precursor to the North West Company (NWC).[14] Rivalries among the Canadian concerns encouraged expansion, but the brazen approach that stiff competition always seemed to encourage delayed Canadian expansion toward the northwestern plains.

The Canadians' aggressive tactics were probably also counterproductive elsewhere, but especially on the plains. Plains bands were much larger, more accustomed to intense and continual warfare, and less reliant on European goods than were bands in the subarctic. On the plains Canadians found themselves on the losing end of confrontations. HBC journals note that Indians killed three Canadians on the Saskatchewan River in 1776–77, dissuading the Canadians from traveling above Cumberland House the following year.[15]

The HBC enjoyed more genial relations with Indians of the plains. This gave them an important advantage. In 1778 Robert Longmoor, halted by

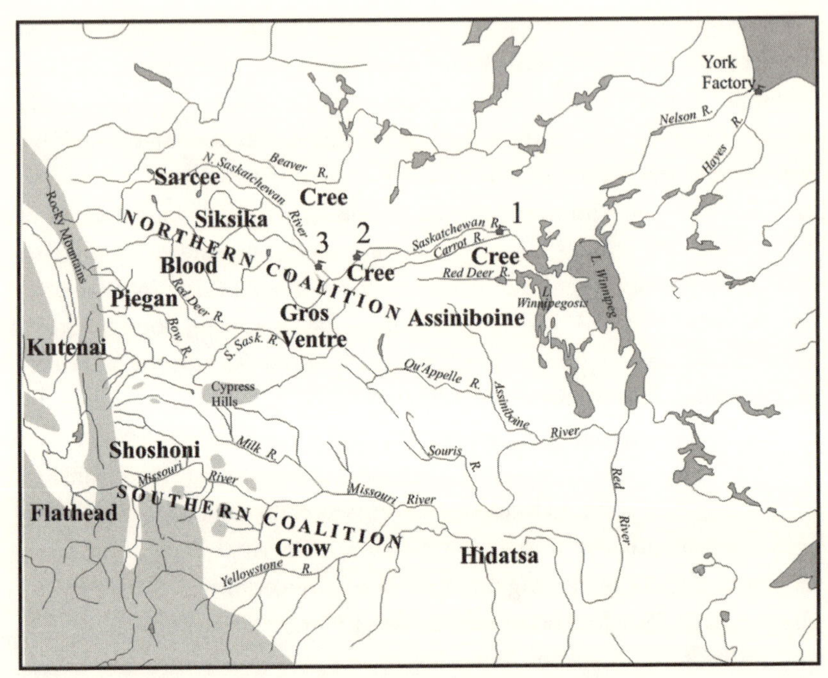

Fig. 6.1. Location of Euroamerican Trading Centers and Indian Groups, about 1781. Between 1774 and 1782 Euroamericans rapidly expanded their operations in the Saskatchewan River Basin. The lower settlements were located in the subarctic forests. They included the HBC's Cumberland House, which was established in 1774, after Canadian traders had already settled there. The lower settlements became an important supply post for the northern brigades. The middle settlements, including the HBC's Hudson House, were established beginning in 1776. They were located near the lowest point on the North Saskatchewan River, from which the bison herds were easily accessible. The upper settlements included Umfreville's House and Pangman's House. They became important supply centers for the bands of the northwestern plains. The fur traders often referred to all posts in the middle settlements and above as the "Forts de Prairies." The entire northern coalition of bands had reliable access to Euroamerican goods, but the southern coalition did not. A broad contested zone separated the northern and southern bands. The locations of ethnic groups shown on this map represent their locations during a typical winter. In reality band territories were not firm. For example, a Sarcee band could spend part of a winter with a Cree band near the Eagle Hills, while a Cree band could winter with a Piegan band near the Bow River. Legend: 1. lower settlements; 2. middle settlements (1776–); 3. upper settlements (1779–).

an unusually early winter, established a temporary HBC post at the Canadians' "middle settlement" at Sturgeon River.[16] The more permanent Hudson House was established fourteen miles downstream the following year. But the Canadians were much farther upstream. Their "upper settlements" were on the lower North Saskatchewan River near the Eagle Hills. On 22 April 1779, however, a group of Crees on the North Saskatchewan River, resentful of the callous, fraudulent, and abusive treatment they received at the hands of Canadian traders, attacked Canadian traders at the Eagle Hills, killing two men, including New Englander John Cole.[17] As a result the Canadians fled their Eagle Hills establishment and left the HBC's Hudson House as the uppermost trading post during the 1780–81 season.[18]

Indians and newcomers became more familiar with one another as a result of the Euroamerican expansion. Until the 1770s the HBC traders identified non-Cree and non-Assiniboine inhabitants of the plains as "Archithinue," but in December 1772 Matthew Cocking elaborated upon these people. After meeting a band of "Yeachithinnee" Cocking explained:

> These Natives are called Powestick Athinnewock or Water-fall Indians. The [Cree] people I am with inform me there are four Nations more which go under the name of Yeachithinnee Indians with whom they are in friendship. Viz. Mithco Athinneewock or Blood Indians, Koskiketow Wathussituck or black foot Indians; Pigonew Athinnewock or muddy Water Indians and Sussewuck or woody Country Indians [Sarcees]. Their Enemies also go under the general Name of Yeachithinnee Indians, four Nations, Kanapick Athinneewock or Snake Indians [Shoshonis]: Wahtee or vault Indians [Hidatsas and perhaps Crows] Kuttunnayewuck [Kutenais] and Nah-puck Ushquanuck or flat Head Indians so called they tell me from their foreheads being very flat.[19]

Thus Cocking's journal for the first time identifies in detail the ethnic groups on the northwestern plains in this era, including groups like the Sarcees who were relatively new to the region. What would be the consequences of Euroamerican expansion for all of these people?

Early in the twentieth century historians agreed that the expansion of trade onto the margins of the northern plains in the last quarter of the eighteenth century caused tremendous cultural changes and economic dependency among the indigenous inhabitants of the region. In contrast anthropologists neglected the effects of Euroamerican traders until pioneer ethnohistorian Oscar Lewis argued that "the fur trade was the mainspring of Blackfoot culture change."[20] According to the prevailing view at the time, the superior technology of the newcomers overwhelmed the Indians, leading them to abandon ancient ways of life. In 1967 E. E. Rich argued that Indians living in the shadow of the Rocky Mountains became dependent on European goods long before Europeans themselves began trading with them directly. He wrote of Indians living in the Hudson Bay basin that

> within a decade of their becoming acquainted with European goods, tribe after tribe became utterly dependent on regular European supplies. The bow and arrow went out of use, and the Indian starved if he did not own a serviceable gun, powder, and shot; and in his tribal wars he was even more dependent on European arms. Steel traps replaced wooden ones more slowly, but by 1743 it was reported that the Indians who traded to York Fort were completely dependent on the arrival of a ship from England.[21]

Beginning in the 1970s historians and ethnohistorians challenged the view that Indians were merely passive and naive victims of unscrupulous traders. Most researchers now believe that Indians were active and sophisticated participants in this trade.[22] The arrival of Euroamerican traders and their wares did not bring rapid economic dependence or cultural and social disintegration but did bring profound change. Every community faced new opportunities and challenges. After 1766 access to European goods was tremendously and increasingly important for Indians on the northwestern plains.

These rapid and substantial changes did not necessarily undermine fundamental aspects of community life on the Great Plains. Just as the immense changes brought on by the horse could strengthen Blackfoot beliefs, so could the coming of European goods and the Euroamericans themselves. The Blackfoot word for Euroamericans, Napikwan (Oldman Person), invokes the image of Napi (Oldman), the Blackfoot creator. The arrival of Europeans evidently reinforced Blackfoot religious beliefs rather than undermined them. Over time the Blackfoot people and other groups understood that Euroamericans were not, as they had initially believed, connected with any "gods."[23] Studies published since 1970 also show that Euroamericans exerted little influence over Indians and that Euroamerican trade goods eased Indian subsistence without transforming the way indigenous people perceived their world. Still, the advantages that access to European goods gave to Indian peoples made it crucially important for groups on the plains to acquire them. It is the unequal access to these goods that acted so significantly to shape the history of the region after 1766.

Significant continuity and rapid change coincided as Euroamerican traders moved inland. The arrival of Euroamerican weaponry among western Cree and Blackfoot bands in the 1730s had already ended Shoshoni-Crow dominance of the northwestern plains by the 1750s. At the same time the effects of the horse and gun revolution encouraged Cree, Assiniboine, and Blackfoot bands to form or strengthen symbiotic relationships in which antagonisms were deliberately soothed. In the late 1760s and the 1770s easier access to Euroamerican goods for Cree, Blackfoot, and Assiniboine bands allowed the northern bands to dominate the northwestern plains as fully as the Shoshoni-Crow system had dominated it in the fifty years before.

The disparity in gun ownership on the northwestern plains, a reality since 1730, intensified between 1766 and 1782. Groups that benefited from the Saskatchewan River trade before 1756 benefited even more as the volume of European goods increased after 1766. In contrast groups that had little access to European goods before 1760 saw no improvement before 1785. During the French era Assiniboine intermediaries dominated the infrequent and desultory trade in European goods at the Mandan and Hidatsa villages. There is no evidence that the Crows or Shoshonis acquired firearms via the Missouri villages at that time. Since there is little evidence of any direct trade between these villagers and Euroamerican traders again before 1785, it is unlikely that the Crows or Shoshonis had access to European goods via the Missouri villages in those years either.[24] Their military position on the northwestern plains deteriorated after 1766.

The Blackfoot, Gros Ventre, and Sarcee bands traded directly with Euroamericans infrequently between 1766 and 1782, but their supply of European goods via Cree and Assiniboine intermediaries grew larger and more secure. The volume of goods transported to the northwestern plains increased substantially. These are evidently the years when many Indians of the region turned to European goods to replace handmade articles. The archaeological record shows that the Blackfoot people had produced pottery for many years. The last documentary reference to Blackfoot pottery and fire-making implements, however, dates to 1772.[25] Indians also adopted metal arrowheads very readily.

Just as their subsistence became easier, the military power of the northern coalition grew. The unequal distribution of European weaponry allowed the northern coalition, including Cree and Assiniboine participants, to wage unrelenting warfare on the southern coalition,[26] who had scant access to European goods and found themselves increasingly vulnerable. The Blackfoot bands, who had abandoned the Oldman River basin only decades earlier, yearly fought to regain access to the rich resources of that region. The still horse-poor Cree and Assiniboine bands joined in the warfare against the horse-rich southern groups. Only these nomadic groups' inability to control and patrol the vast region allowed others to maintain themselves there. Shoshoni, Crow, Kutenai, and Flathead bands continued to live in the upper Missouri and Oldman River basins, but their battles were primarily defensive.

While the coming of Euroamericans intensified some old patterns, it also reshaped patterns of trade, warfare, and diplomacy. Only some relationships changed manifestly before 1782, but the roots of other changes lie in these years. The westward spread of Euroamerican traders on the Saskatchewan River undermined established patterns of indigenous trade, with enormous and enduring implications for northern plains Indians. As trading establishments proliferated, the Cree bands of the Saskatchewan River area lost the ability to profit from their role as intermediary traders. Still, most Cree bands seem to have welcomed the traders.

Until the 1750s the western woods Cree and the Assiniboine bands were only minor players on the northwestern plains.[27] Some Assiniboines resided on the northeastern plains, while the Crees were primarily forest and parkland dwellers toward the north and east. After the Crees and Assiniboines acquired horses in the 1750s, however, some gravitated increasingly to the northwestern plains. The move from parkland to plains required lifestyle changes. Canoes served parkland and forest communities very well and were essential for trading expeditions to Hudson Bay, but they did not suit plains bison hunters and were unnecessary for visits to trading posts on the North Saskatchewan River.

In the 1750s and 1760s the rise of equestrianism enticed those trading bands that had regular contact and cordial relations with the Blackfoot and Gros Ventre bands to adopt a bison-hunting plains existence. As they adopted this lifestyle, individuals and bands moved to the west and south. In 1787 David Thompson told Young Man that "sons of those he left there [at the Pasquiau River] hunted on the north bank of the [North Saskatchewan] River, many days march above it, that the lowest of them were on the west side of the Eagle Hills and that his country was now hunted upon by the Indians whom in his time were eastward of Lake Winipeg."[28] The Crees and Assiniboines were able to move onto the northwestern plains without a struggle because they had cordial relations and kin connections with plains bands who not only accepted them but assisted their transition.

The transition was slow until European traders arrived in the region in the late 1760s, 1770s, and 1780s. Even in 1780 William Tomison met four Sturgeon River Crees who had been at a bison pound, but "they say they have had but little Success as yet by reason they are all Southward Indians, that is there, and not thoroughly acquainted with the method of driving Buffalo into the pound."[29] It was the need to make the annual canoe trip to Hudson Bay that hampered the adaptation to a full-time equestrian and bison-hunting lifestyle. As the Blackfoot traders had already understood, the annual trip to the Hudson Bay was uninviting to equestrian bands. So when Euroamerican traders arrived on the Saskatchewan River the trading bands who had relied on their role as intermediaries welcomed their arrival because it facilitated their adjustment to a plains lifestyle without jeopardizing their access to European goods.[30]

It is also important to understand that the return of Euroamericans to the Saskatchewan River after 1763 was not a mere reenactment of the earlier French trade. The number of non-Indians on the Saskatchewan River quickly exceeded the number during the French period. Furthermore, when traders pressed on into the northwestern subarctic forests where the finest furs on the continent were to be found they knew that they would need a new source of provisions. The northern plains met this need well, because it was near the major supply routes through the subarctic.[31]

Furs were plentiful in the lower Saskatchewan River region, but large game was scarce there, as in much of the subarctic. Almost immediately after they established posts along the lower Saskatchewan, the traders turned to the upper Saskatchewan, to acquire not only more furs but also provisions.[32] Canadians cultivated a provisions trade by the early 1770s, and the HBC did so later in the same decade. When the HBC built Cumberland House in 1774, Canadian traders were already established considerably farther upstream. The traders at Cumberland House, however, soon discovered what earlier traders must have learned quickly and what local Indians warned them of: the resources of the region could not support as many men as they wished to keep at the post. This led inevitably to the conclusion reached in early 1777 that

> this inforces the necessity of the Company's making an early Settlement up above towards the Buffalo Country, where men may most likely be well provided for. . . . Provisions may also be Collected at an Upper Settlement to assist this Place; Since the Pedlers have been so numerous, those who resided at the Upper Settlements have generally provided a Supply of Provisions to help their fellow Traders in the Spring who resided in the Lakes, otherwise these would be distressed in their Journey Down.[33]

Accordingly, when Robert Longmoor built Hudson House in 1778 he located it near the Thickwood Hills to the northwest and the prairies to the south to serve the dual purpose of collecting moose and bison meat and valuable furs. The post was on the border between plains and parkland at the lowest point on the North Saskatchewan, from which the bison herds were easily accessible.

The HBC, however, was a relatively small buyer of plains provisions. In 1776 there were nearly a hundred Canadian traders on the Saskatchewan River. Only two years later the Canadians numbered about three hundred.[34] Unlike previous Indian traders, who were subsistence hunters on trading expeditions, the Euroamerican newcomers were trading specialists who kept their subsistence efforts to a minimum. When they traveled, Canadian traders needed to minimize cargo space devoted to provisions and time devoted to hunting. Saskatchewan River traders themselves sought to buy provisions in the late 1760s and early 1770s, but the provisions market grew tremendously in the late 1770s and early 1780s, when Canadian traders expanded their operations into the rich fur lands of the north.

The subarctic forests simply did not have the resource base to feed the traders. They needed a compact, nutritious, and imperishable food supply to take with them. Pemmican was the answer. Traders on the northern plains bought fresh, dried, and pounded bison meat that post personnel rendered into pemmican and transported downstream to Cumberland House, the main provisions supply depot for northern canoe brigades. Especially for many Cree and Assiniboine bands on the northern plains, the provisions trade became important after 1766.

The provisions trade further encouraged Cree and Assiniboine moves toward the plains. Cree and Assiniboine bands that formerly lived on the plains only seasonally as traders now gravitated toward the plains full-time, primarily as provisioners. To enhance their position, they hindered traders' efforts to hunt for themselves. They used an old tool: fire. By 1776 traders reported that Cree bands who were increasingly trading provisions rather than furs deliberately burned grass near trading posts in a successful attempt to drive up the price they could get for meat. In the fall of 1780 HBC traders noted that the "Natives have burnt all the Ground, that nothing can Stay on it, their design is that they may get a great deal for provisions, as a very few is hunting of furs."[35]

The approach and arrival of Euroamericans in the 1760s and 1770s was a separate process from the horse and gun revolution that had preceded it. In many ways the actual arrival of Euroamericans reinforced the patterns established earlier, but in other ways it brought important changes. The smallpox epidemic of 1781–82 compounded these changes, intensifying established patterns of human interaction but also unleashing new forces.

After sparing the residents of Mexico City for almost twenty years, a particularly virulent strain of smallpox struck in August 1779, gaining a solid foothold in the city the following month. The disease raged into early 1780, killing about 18,000 residents.[36] It also spread to New Mexico as early as 1780, devastating the isolated New Mexicans just after they had endured the worst decade of Indian raids in the history of the colony. Comanche, Apache, Pawnee, Wichita, and Osage warriors raided New Mexico relentlessly during the 1770s.[37] Juan Bautista de Anza finally defeated one of the most prominent Comanche war leaders, "Cuerno Verde," in 1779.[38] In 1780, either in an attack or in a peaceful encounter with the New Mexicans, Indian communities contracted smallpox, touching off a pandemic that reached as far as the Chipewyans of the Hudson's Bay coast by 1783.

If smallpox ever hit the northwestern plains before 1780, there is scant evidence of it. In 1781 horses and the frequent warfare that accompanied them almost certainly helped spread the disease north along the important horse-diffusion routes between New Mexico and the northwestern plains and the Mandan-Hidatsa villages.[39] Northern Indian groups frequently traveled to the Spanish settlements to trade for European goods and to raid horses. Once they contracted the disease they were asymptomatic for ten to fourteen days—ample time for these highly mobile people to spread the disease a long way.

Smallpox probably hit the northwestern plains in the spring or summer of 1781, striking southern bands first. The disease spread to the northern coalition from the Shoshonis when a combined Cree, Assiniboine, and Blackfoot war party (possibly including Gros Ventres) attacked a camp of dying Shoshonis. In 1782 Matthew Cocking wrote that the

> Southern, Assinnee Poet, and the Yachithinue met with a tent of Kanasick Athinewock (i.e.) Snake Indians who were all ill of the Small Pox (and where supposed to have recieved from the Spaniards whom tis said those people trade with) killed them all and scalped them to carry away with them, by this means they received the infection and almost all of them died on their return, what few reached their own country communicated the disorder to their Friends and it spread through the whole country above here in some parts of which it still rages.[40]

Young Man's description is the most powerful account of that pandemic. Warriors of the northern coalition first caught the disease, apparently in the spring or early summer in the Red Deer or Bow River region:

> We caught it from the Snake Indians. Our Scouts were out for our security, when some returned and informed us of a considerable camp which was too large to attack and something was very suspicious about it; from a high knowl they had a good view of the camp, but saw none of the men hunting, or going about; there were a few Horses, but no one came to them, and a herd of Bisons [was] feeding close to the camp with other herds near. This somewhat alarmed us as a stratagem of War; and our Warriors thought this camp had a larger not far off; so that if this camp was attacked which was

strong enough to offer a desperate resistance, the other would come to their assistance and overpower us as had been once done by them, and in which we lost many of our men.

The council ordered the Scouts to return and go beyond this camp, and be sure there was no other. In the mean time we advanced our camp; The scouts returned and said no other tents were near, and the camp appeared in the same state as before. Our Scouts had been going too much about their camp and were seen, they expected what would follow, and all that could walk, as soon as night came on, went away. Next morning at the dawn of day, we attacked the Tents, and with our sharp flat daggers and knives, cut through the tents and entered for the fight; but our war whoop instantly stopt, our eyes were appalled with terror; there was no one to fight with but the dead and the dying, each a mass of corruption. We did not touch them, but left the tents, and held a council on what was to be done. We all thought the Bad Spirit had made himself master of the camp and destroyed them. It was agreed to take some of the best of the tents, and any other plunder that was clean and good, which we did, and also took away the few Horses they had, and returned to our camp.

The second day after this dreadful disease broke out in our camp, and spread from one tent to another as if the Bad Spirit carried it. We had no belief that one Man could give it to another, any more than a wounded Man could give his wound to another. We did not suffer so much as those that were near the river, into which they rushed and died. We had only a little brook, and about one third of us died, but in some of the other camps there were tents in which every one died.[41]

Traders learned about the disease only in the fall. On 22 October 1781 a Cree man who had recovered from smallpox arrived to tell the HBC trader at Hudson House, William Walker, about the devastation the disease had wrought. On 11 December William Tomison learned of it at Cumberland House.[42] The disease also spread to the northern plains via the Mandans and Hidatsas. They were so devastated by the epidemic that the villagers at Heart River and Knife River, who numbered around 10,000 in 1780, resettled into smaller villages at Knife River after the epidemic.[43]

The smallpox epidemic affected community relations on the northwestern plains in part because of its uneven patterns of mortality. The epidemic apparently hit all the indigenous groups there but affected very few Euroamerican newcomers. It is difficult to estimate how many Indians of the northwestern plains died in the 1781 epidemic. Traders' estimates of the death rates among the Indians of the Saskatchewan River may not reflect those on the northwestern plains.

Traders' reports themselves suggest that the toll on the plains was generally lower than in the forests. After learning of the many deaths at Cumberland House, William Walker at Hudson House wrote that "there is a good few Indians alive Up here Yet."[44] Timing was on the side of the Indians of the northwestern plains. The epidemic hit during the warmer months, when its victims were less vulnerable to death by exposure and starvation. In contrast the Indians of the Saskatchewan region suffered during an exceptionally cold spell during a winter of unusual scarcity. The traders' most detailed initial reports were exaggerated. They discovered later that fewer Indians had died than they first guessed. For example, Walker at first incorrectly assumed that the entire Touchwood Hills Assiniboine band had died.[45] Young Man's explanation that the disease killed every member of some plains bands but only a third of others is useful. Overall the disease probably killed about two-thirds of all northwestern plains Indians.

The varying death rates among different Indian bands and between Indians and Euroamericans were not caused by genetics but by particular historical circumstances. If the Indians of the northwestern plains were exposed to smallpox before 1780, they evidently had no memory of it. None of them could have acquired an immunity from previous exposure to the disease. Under these conditions the disease attacked entire communities, young and old, simultaneously, leaving no healthy members to care for the sick. Unable to follow game or gather firewood and water, such communities faced not only the primary effects of the disease but also a combination of starvation, exposure, and dehydration. And so the death toll climbed. William Tomison explained how some bands, forced to move to find food, were compelled to abandon sick relatives to an almost certain death.[46] In sedentary Indian communities elsewhere in North America crowded and unhygienic living conditions similarly boosted mortality. In Europe, where smallpox was common, only a small number of people were ill at any time. Healthy caregivers could often nurse the sick to health. Euroamericans' sedentary lifestyles, stored food reserves, and relief institutions were an advantage to the sick, even in those communities (especially in North America) where smallpox hit an entire community at once.

Indians also died in very high numbers because, without any experience with the disease, they often responded inappropriately. Europeans had been treating smallpox since the sixteenth century. For many years they had experimented with and abandoned various treatments that only increased mortality.[47] In 1781 the Indians of the northwestern plains used treatments that were often as counterproductive as those Europeans had earlier used. Traders did not have to worry about treating their own sick because few even contracted smallpox. Only one mixed-blood HBC employee is recorded as having contracted the disease, and he recovered. William Tomison mused that "there is something very malignant, that we are not sensible of, either in the Constitution of the Natives or in the Disorder, those that Die before the smallpox breaks out is tormented with great pains and many of them Die within 48 Hours."[48]

In reality smallpox reached its height in mortality in Europe in the eighteenth century.[49] The most conservative recent estimates suggest that in the eighteenth century smallpox caused ten percent of all deaths in Europe, and only the introduction of the cowpox vaccine in the nineteenth century significantly reduced this mortality.[50] The disease carried off men and women of all ranks of society from the meanest peasant to King Louis XV of France, who died of smallpox seven years before the epidemic hit the northwestern plains.[51] In contrast to the situation in North America, however, smallpox was endemic in large cities like London, and epidemics of the disease passed through rural English towns and the countryside approximately every four or five years. Smallpox in England in the eighteenth century was a childhood disease that usually struck people by the time they were five.[52] For the rest of their lives the survivors were immune to the disease.

This explains why none of the HBC's employees from England contracted the disease in 1781. Most of the HBC's inland employees in 1780, however, were from the Orkney Islands, off the northern coast of Scotland.[53] Smallpox could be absent for decades on those northern islands.[54] When it did hit, it tended to strike most of the population, with devastating results. It is unclear whether there had been epidemics on the Orkney

Islands that rendered the HBC employees immune to smallpox in 1781. All of the HBC's Orcadian servants may have been immune in any case. Because the disease was more devastating in outlying communities than in London or even in rural England, and because many people from these communities were seafarers, Orcadians accepted smallpox inoculation more quickly than people in major population centers. Orcadians carried out the first recorded general inoculation for smallpox in Scotland in 1783, and only a small number of people were inoculated in that campaign because inoculation had been extensive on the island even before that.[55]

Although historians credit Lady Mary Wortley Montagu, wife of the British ambassador to Turkey, with popularizing inoculation among the elite of England after 1721, rural Europeans seem to have been inoculating themselves before that. Evidence of inoculation in Scotland dates as far back as 1715.[56] Since lay practitioners and even parents sometimes inoculated children,[57] the practice may have been more popular than official records indicate. The documentation regarding the mortality of Canadian traders is fragmentary, but it suggests that smallpox killed some Canadian traders in the interior. It is evident that European adults on the northern plains were not genetically less susceptible to smallpox than Indians were; like most adults in Europe in the eighteenth century, they had acquired immunity through contact with the disease as children.

Differing death rates among Indian bands and coalitions must have changed patterns of human interaction among the communities in ways that we cannot know. The fact that the Euroamericans suffered virtually no losses from the disease certainly affected their role in the region. After 1781 the Euroamerican trade expanded rapidly into the upper North Saskatchewan River basin. The devastation of the epidemic may have made the Indians more willing to welcome traders to the region, but it is unlikely to have been a significant factor. The rapid expansion of trade would certainly have been impossible, however, if the traders had lost experienced leaders in the same proportion as Indian groups did.

The smallpox epidemic of 1781 was an immeasurable catastrophe. The demographic collapse represents only one aspect. The epidemic must have so dramatically affected American Indian communities that most, if not all, bands on the northwestern plains ceased to exist as autonomous units. In the weeks and months following the epidemic, decimated bands, many of whom lost their most prominent leaders, must have merged with other bands to form new communities. The documentary evidence does not describe this process among the Indians of the northwestern plains, although Larocque in 1805 hinted at it for the Crows, writing that "since the great decrease of their numbers they generally dwell all together, and flit at the same time as it is possible for them to live when together, they seldom part."[58] Documentation of the change in band structures is more complete for the Crees of the Saskatchewan River region.[59] While the exact implications can never be known, it is safe to assume that formerly separate bands amalgamated in the months following the epidemic. The merging process was inevitably accompanied by social strains and instability and was probably as important as the demographic collapse itself.

Certainly epidemics that killed at least a third of almost every band on the northern plains had ramifications beyond demographics. Today few scholars accept Calvin Martin's explanation of the implications of epidemics for indigenous worldviews, but this remains one of the few studies that address the question.[60] We must consider not only what portion of the people in Indian communities were killed in epidemics but what role these victims had in their societies. Alfred Crosby has noted unsentimentally that endemic diseases tend to kill children, "the most expendable and easily replaced members of society."[61] Epidemic diseases kill differently. Indian bands in 1781 lost a cross-section of their population. Survivors mourned loved ones while they recovered from their own traumatic experience. Many carried permanent and disfiguring scars as constant reminders. Indigenous societies tended to lose their least expendable and least replaceable members during smallpox epidemics. Adult males apparently died in disproportionately high numbers, both in the northern forests and on the plains.[62] The impact on Indian communities must have been tremendous.

What psychological, emotional, and spiritual effects did the smallpox epidemic have? The scant evidence only suggests the enormity of the event. About six years after the epidemic Young Man told David Thompson:

> When at length it [the smallpox] left us, and we moved about to find our people, it was no longer with the song and the dance; but with tears, shrieks, and howlings of despair for those who would never return to us. War was no longer thought of, and we had enough to do to hunt and make provision for our families, for in our sickness

> we had consumed all our dried provisions; but the Bison and Red Deer were also gone, we did not see one half of what was before, whither they had gone we could not tell, we believed the Good Spirit had forsaken us, and allowed the Bad Spirit to become our Master. What little we could spare we offered to the Bad Spirit to let us alone and go to our enemies. To the Good Spirit we offered feathers, branches of trees, and sweet smelling grass. Our hearts were low and dejected, and we shall never be again the same people. To hunt for our families was our sole occupation and kill Beavers, Wolves and Foxes to trade our necessaries; and we thought of War no more, and perhaps would have made peace with them [the Snakes] for they had suffered dreadfully as well as us and had left all this fine country of the Bow River to us.[63]

Young Man's poignant account hints at the far-reaching effects of the epidemic. Communities appear to have turned inward as they worried more about their day-to-day existence than about their relations with outsiders. They turned from war to subsistence. Young Man suggests that the epidemic even shook the Blackfoot people spiritually.

Another story also hints at the effects of the epidemic. When David Thompson saw the remains of the "One Pine" during the winter of 1787–88, his Piegan hosts told him:

> This had been a fine stately tree of two fathoms girth, growing among a patch of Aspins, and being all alone, without any other pines for more than a hundred miles, had been regarded with superstitious reverence. When the small pox came, a few tents of Peeagans were camping near it, in the distress of this sickness, the master of one of the tents applied his prayers to it, to save the lives of himself and family, burned sweet grass and offered upon its roots, three horses to be at it's service, all he had, the next day the furniture of his horses with his Bow and Quiver of Arrows, and the third morning, having nothing more, a Bowl of Water. The disease was now on himself and he had to lie down. Of his large family only himself, one of his wives, and a Boy survived. As soon as he acquired strength he took his horses, and all his other offerings from the "Pine Tree," then putting his Axe in his belt, he ascended the Pine Tree to about two thirds of it's height, and there cut it off, out of revenge for not having

> saved his family; when we passed the branches were withered and the tree going to decay.[64]

In a different version of this story Thompson noted that the man believed that this tree "had been planted by the evil spirit." When Thompson met this man during the winter of 1787–88 he was "a good looking person with a deep settled melancholy upon his countenance." Young Man explained the man's story: "after the loss of his family, he had taken no wife, lived alone in the tent of one of his brothers; that he had been several times with the war parties, never took a shield with him, always placed himself in the front of the battle as if he wished to die and yet none of the enemies arrows ever struck him."[65]

Communities long remembered the epidemic of 1781. In 1876 the Crows gave James Bradley a vivid account:

> Something less than a hundred years ago the Crows were living in two bands, the greater portion making their home upon the waters of the Powder River, while the smaller band of four hundred lodges, or about four thousand souls, were camped in the lower extremity of the Clark's Fork bottom, along the base of these bluffs. Here a terrible disease broke out among them, the victims being covered from head to foot with grievous sores. It proved fatal and destroyed almost the entire band. The plain was covered with the bodies of the dead, and their horses ran wild because there was no one to take care of them. The few who escaped the disease fled to the village in Powder River.[66]

When so many people died at once, communities inevitably lost important repositories of community knowledge. David Thompson and other traders noted that both on the plains and in the forests game seemed scarce after the epidemic.[67] The scarcity probably had as much to do with perceptions as with reality. If adult males died in disproportionate numbers, the community would have been less able to predict the movements of game and to hunt effectively, creating food shortages.

Although there is no evidence showing that certain groups on the northwestern plains suffered greater mortality than others, some groups do appear to have suffered more in military terms. Groups afflicted first were unaware for a time that their enemies would also be struck, so they often withdrew from contested grounds. That, according to Young Man, is how the Shoshonis fled the Bow River basin after the epidemic.[68]

The epidemic of 1781 brought the end of the Plains Kutenais. In 1811 Alexander Henry the Younger explained that he understood (probably from information given by Blackfoot or Cree informants) that the Kutenais, "being driven into the Mountains by the different tribes who inhabited the Country to the Eastward of them, and with whom they were perpetually at War, they in their turn waged War upon their harmless neighbours to the Westward, the Snare Indians, and soon drove them away from off the Lands the Kootonaes now inhabit, which is the upper part of the Kootonaes or Columbia River."[69] The Kutenais still hunted on the plains seasonally, but it seems that none of them lived east of the Rocky Mountains permanently after 1781. The Kutenais' gradual retreat was fundamentally caused by their enemies' attacks, but the smallpox epidemic probably prompted the final withdrawal in 1781.[70] The epidemic weakened the position of the Shoshoni bands and their neighbors in comparison to the Blackfoot bands and their neighbors. Traders on the Saskatchewan River were willing to pay higher prices for provisions, at least temporarily, after 1781, further strengthening the position of the northern coalition.[71]

The epidemic of 1781 left every indigenous community on the Great Plains reeling. Never has so large a portion of the population of the plains faced such a calamity. Never have so many communities simultaneously faced such a multitude of challenges. The epidemic also affected relations within and among societies, while Euroamerican traders, largely unscathed, continued their expansion westward unabated. Every Indian band in the region probably lost important leaders, adopted new members, or joined other bands. Bands of the southern coalition, who were already vulnerable to attack from their more powerful neighbors, withdrew toward the southwest. Although the bands of the northern coalition faced much of the same devastation as their enemies, their military and territorial position actually improved as a result of the epidemic. Even Young Man, however, believed that his people would be forever changed by the experience. Neither Young Man's people nor any of the other bands on the northwestern plains enjoyed a period of peace and quiet during which they could adjust to the new realities, for they were on the brink of a new era of turbulent change.

CHAPTER SEVEN

"Many Broils and Animosities," 1782–95

The Crees being the most powerful clan in this quarter, have been involved in frequent quarrels with the Gros Ventres for many years past, but as they mutually feared each other their hostilities amounted only to the death of a few of either party, when they occasionally met at the Fort, 'till summer—93.

—DUNCAN MCGILLIVRAY
1795

This morning a Band of about 30 Blood Indians and 10 Black feet arrived. . . . They all confirm the former rumours that the Gros Ventres *are separated in Two Bands, one of which consisting of 90 lodges direct their course towards this quarter, and the other have formed an alliance with the* Snake Indians,*—a tribe who inhabit the Rocky Mountains, unacquainted with the productions of Europe, and strangers to those who convey them to this Country. The Snake Indians have suffered a severe loss in War this year if rumour be true, a party of the two first mentioned tribes having killed no less than 25 men and two Women in a recent expedition against them.*

—DUNCAN MCGILLIVRAY
1795

The years after Euroamerican traders arrived on the northern plains were tumultuous, because every community faced new opportunities and challenges. It became uncertain whether the Crow, Shoshoni, Kutenai, and Flathead bands could continue to live and hunt on the northwestern plains. Meanwhile the northern coalition began to dissolve. As the northern bands considered the possibilities offered them and limitations imposed upon them by the expansion of trade, divisions deepened. Certain bands, chiefly Cree and Assiniboine, still enjoyed privileged access to European goods. Bonds of friendship between these privileged bands and other plains bands loosened, however, as the mutually beneficial relationship that had linked them for at least two generations dissolved. The kinds of antagonistic incidents that had earlier been treated with restraint and toleration began to cultivate animosity, jealousy, and resentment. Occasional raids escalated into skirmishes, skirmishes into battles, and battles into full-scale warfare. Blackfoot and Sarcee friendships with Cree and Assiniboine bands deteriorated but did not dissolve, although by 1794 only remarkable forbearance and diplomacy preserved any sense of common ground. By then Gros Ventre friendships with the Crees and Assiniboines were irreparably damaged. The Gros Ventres seemed to be turning toward the southern bands, with whom they had more in common.

The result was warfare in the Saskatchewan River basin on a scale perhaps never before experienced. Rivalries among traders also intensified. Unable to stem the tide of bitterness, Euroamericans vainly presented themselves as neutral peacemakers to the Indians, naively believing that they would not become enmeshed in the hostilities themselves. They were mistaken. Even when they did become embroiled in Indian hostilities, however, the traders understood that they were minor players caught in the crossfire of a deadly battle in which there could be no immunity for dealers in arms.[1]

Between 1782 and 1794 Euroamerican competition for the furs of the northwestern interior of North America increased, although the HBC held its ground against its Canadian rivals only in the Saskatchewan River basin. Canadians led in exploration and trade farther north. Alexander Mackenzie, for example, reached the mouth of the Mackenzie River in 1789 and became the first Euroamerican to cross the continent of North America in 1793. Explorations propelled traders deeper into the vast subarctic forests of the north.

Peace between Britain and the United States accelerated a process that had begun after hostilities broke out in the late 1770s. War had convinced many involved in the trade of the southwestern region to turn toward the northwest. That is how men like Peter Pond, Peter Pangman, and Alexander Henry came to the area. During the 1780s the trade of Lake Michigan and the Mississippi still was double that of the northwest.[2] Then uncertainties in the new United States, hostilities with Indians in that country, exhaustion of the fur resources of the southwest, and news of the tremendous profits to be earned in the northwest trade drew many more traders toward the northwestern interior. Excluded from the North West Company (NWC), Gregory, McLeod and Company struggled bitterly and violently with its rival between 1783 and 1787. The HBC was far less aggressive. It had solved its transportation problems by 1780, but then Britain entered several decades of almost continuous warfare. The British military's demands ensured that the HBC faced a chronic manpower shortage. Only in the Saskatchewan River basin did the HBC nearly keep pace with the NWC, which increasingly dominated the Montreal-based trade.

As the traders drove into the subarctic forests, they required more plains provisions than ever. The North Saskatchewan River, which flowed primarily through the parkland region along the northern margins of the plains, was convenient because traders there could collect valuable furs in addition to provisions. Traders built few posts on the South Saskatchewan (South Branch) River, which ran through the plains (see figure 7.1). Traders' posts were limited to the northern margins of the northwestern plains. On the North Saskatchewan River rival traders built posts near one another, sometimes within a single palisade, so they could keep a constant watch on their competitors and could work together should Indian bands threaten them. They also built all their posts on the north side of the river or on an island. At least in the summer, when traders felt especially vulnerable, plains bands would camp on the south side of the river and send their traders across on horseback or on boats, with the help of traders, to reach the posts.

Competition among Canadians, especially before the NWC and Gregory, McLeod and Company merged in 1787, continued to impel traders upriver. In the summer of 1785 Canadian traders, who apparently numbered almost two hundred, established a post at the Battle River near the Eagle Hills. The HBC was left behind that year; but in 1786 a Canadian, Donald Mackay, invited William Walker at Hudson House to

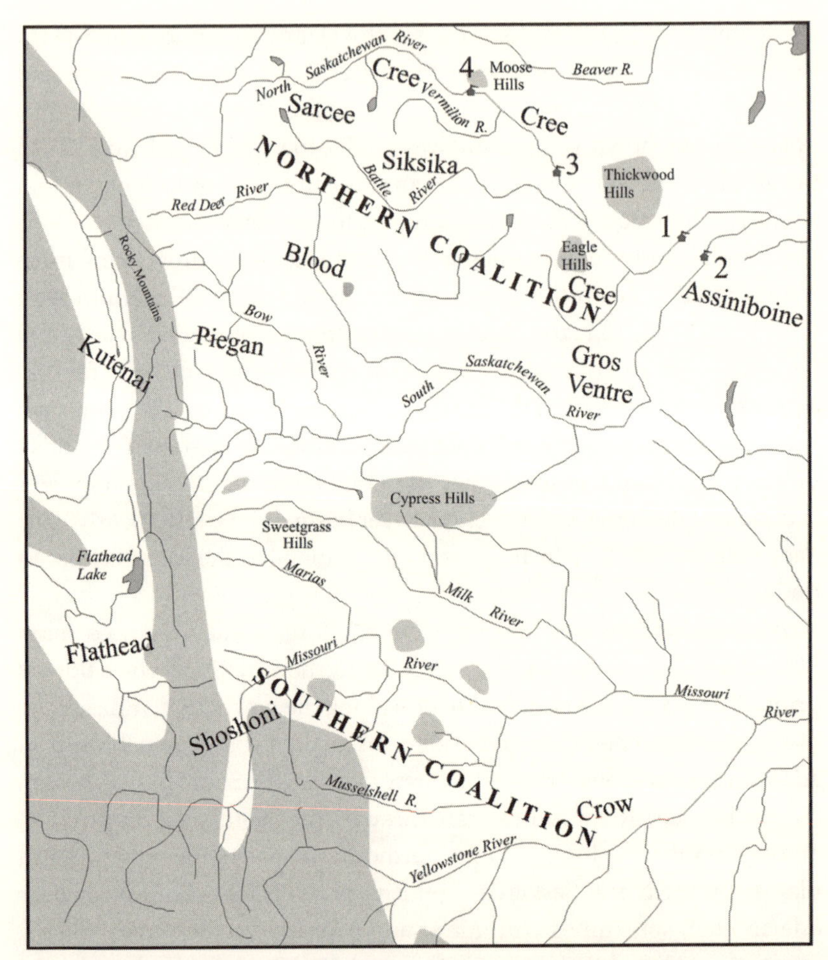

Fig. 7.1. Location of Euroamerican Trading Centers and Indian Groups, 1782–94. Legend: 1. Hudson Houses (Upper and Lower), 1778–88; 2. South Branch Houses, 1785–94; 3. Manchester House / Pine Island Fort (Fort de l'Isle), 1785–94; 4. Buckingham House / Fort George, 1792–1801.

join him in establishing posts about sixty miles beyond the Battle River at a location recommended by Cree or Assiniboine leaders there.[3] Mackay and Walker were joined by Peter Pangman of Gregory, McLeod and Company and William Holmes of the NWC at what is known as the Pine Island / Manchester House complex.[4] In the fall of 1786 local bands could choose from four competing posts at Pine Island and two others: Edward Umfreville's, thirty to forty miles upstream, and another at the Battle

River. William Tomison complained that the Indian "has nothing to do but go from house to house, get Drunk and beg Goods on Expectation of what they are to bring."[5]

The two main Canadian firms also established South Branch House about forty miles up the South Saskatchewan River in 1785. In the autumn of 1786 the HBC built its South Branch House there to replace Hudson House, which it abandoned as a permanent establishment.

After the amalgamation of the NWC and Gregory, McLeod and Company in 1787 the competition for the furs of the Saskatchewan River basin was increasingly a struggle between the reorganized NWC and the HBC. Without significant Canadian rivals, the NWC prospered tremendously between 1787 and 1795.[6] In 1789, when Angus Shaw of the NWC traveled about six days above Manchester House to establish another post, the HBC was unable to answer the challenge.[7] In the spring and autumn of 1792, however, the NWC and HBC built Fort George and Buckingham House, respectively, near the Moose Hills.[8]

The new posts marked the beginning of a new era on the northwestern plains. Hudson House, established in 1778, was within reach of the Gros Ventre, Sarcee, and Blackfoot bands, but the era of regular direct trade really began after 1782. Once Manchester House and Buckingham House had been established, the Gros Ventre, Blackfoot, and Sarcee traders ceased visiting Hudson House.

The first recorded visit of Gros Ventre traders to trading posts occurred when some Gros Ventres visited Hudson House in November 1779, but substantial direct trade began only after 1782. Direct Blackfoot trade with Euroamericans also appears to have begun before 1782, although it was not regular until after that year. The easternmost Blackfoot groups established trading relations first. Initially the NWC captured most of their trade because none of the HBC men learned the Blackfoot language until the mid-1780s. When nine Siksika men visited Hudson House in 1782, most went to the NWC post, leading William Tomison to explain that "we are greatly at a Loss of not knowing the language of this tribe as also the fall Indians, and indeed never will, without Men goes, and lives with them for some Years, as that is the way that the Canadians have acquired the Language."[9]

Tomison began sending envoys to the Blackfoot groups. In May 1784 he "fitted out James Gaddy & sent him away with the Picanau & Blood

Indians, to learn the Language[.] [T]he Indian that I have sent him in care of talks the Southern Tongue well; as also Blackfoot; & seems to be a *good sort of a Man*. I have also sent Isaac Batt to meet the Blackfoot Indians & endeavour to bring them in with Provisions in the Sumr." Visits soon became routine. By the autumn of 1786 Batt and Gaddy were among four people Longmoor sent from the newly established Manchester House to winter with the Piegans. In 1787 David Thompson was one of eight men sent to the Piegans, while another six went to the Bloods.[10] The Piegans may have been the last of the Blackfoot groups to establish trading relations with Euroamerican traders (evidently there were still Piegan leaders who were visiting posts for their first time in 1794), but the traders soon particularly sought the Piegan trade.[11] They were the only Blackfoot group to bring substantial numbers of beaver furs to trade, although they shot their beavers during the warmer seasons when their fur was not as valuable. Even then some refused to hunt or even touch beavers for personal religious reasons.[12]

The Sarcees apparently first visited Hudson House in 1783, when Tomison noted the arrival of a tent of "Sussuwich," "Sussu," or "Pelican Indians" that "talk a different Language from either fall or blackfoot Indians." Their visit was not a perfect success. After leaving the post the members of this small group met a large Cree or Assiniboine band that forced them to give up some of what they had acquired at Hudson House. In 1786 the HBC sent its first envoy to live with the Sarcees for a winter.[13] Based on his experiences ending in 1788, Umfreville said that the Sarcees "now harbour in some country about the Stony Mountain, where they keep to themselves, for not many have as yet appeared at any of the trading houses."[14]

Hudson House was evidently outside the territory of every Archithinue band, for the Blackfoot traders ceased visiting there after 1785, when Canadian traders built posts farther upriver. Thereafter Hudson House and South Branch House, its replacement, were Cree and Assiniboine posts. When Manchester House and Pine Island were established, all of the Archithinue traders preferred trading there. During those posts' first years the Blood and Piegan traders provided most of the furs brought to Manchester House.[15] The Sarcees also visited Manchester House often in the 1787–88 season. When the Buckingham House/Fort George complex was built, some Gros Ventre and Siksika traders continued visiting

Manchester House, but most of the Blackfoot and Sarcee traders resorted to the more convenient Moose Hills posts, as did some Assiniboine and Cree bands.

After an inauspicious beginning the HBC appeared to be capturing a growing portion of the plains trade during the 1780s. Canadians placed greater value on the trade of other regions, and their aggressive tactics were ineffective with the plains bands. Even into the 1790s the NWC had more Blackfoot interpreters than the HBC, but it seemed unable to use this advantage to attract customers. In 1787 William Tomison noted that traders at Peter Pangman's post had locked up a party of Gros Ventres and forced them to trade with them. In 1788 he was satisfied to see that, although seven Canadians had accompanied a band of Piegans all winter, all but one man in the band traded at his post.[16]

Other groups on the northwestern plains met Euroamerican traders for the first time during the 1790s, but it did them little good. In the fall of 1792, while visiting Sakatow's Piegan band to negotiate peace agreements, some Shoshonis and Kutenais met two HBC traders, but the Piegans prevented any trade. Peter Fidler explained that the Kutenais had never visited trading posts, "altho they much wish it—But the muddy river, Blood, Black Foot, & southern Indians always prevents them—they wishing to monopolize all their skins to themselves—which they do giving the Poor Indians only a meere trifle for—they scarce give them as much for 10 skins as they can get for one at the Trading Settlements." Fidler himself watched as the Piegans "began to barter for horses as soon as we arrived & soon bought all the Cottonahaws had to spare—for a mere trifle some only giving an old Hatchet."[17] Despite the prices for these goods (which did not include guns), the Kutenais traded every horse they had. Their women had to carry all their possessions on their backs.

The Kutenais could not break the northern coalition's blockade. At Fort George on 22 February 1795 Duncan McGillivray reported that "the Coutonées have already made several attempts to visit us, but they have been always obstructed by their enemies and forced to relinquish their design with loss."[18] Into the 1790s the Kutenais, Shoshonis, Flatheads, and Crows remained utterly dependent upon the parsimonious Blackfoot traders and on other tenuous sources of European goods.

Not surprisingly, the northern coalition became more dominant between 1782 and 1794. Manchester House and Buckingham House, established on the very fringes of Blackfoot territory, greatly facilitated the spread of European goods to the plains. The volume and range of goods available increased, and the prices that the traders demanded for new items were a fraction of the prices Cree and Assiniboine intermediaries had asked. While the Blackfoot traders had acquired used guns from Cree and Assiniboine intermediaries at 50 MB, they bought them for 14 MB at the Saskatchewan River posts in 1794.[19]

The Crows, Shoshonis, Flatheads, and Kutenais also must have gotten more European goods in these years, no thanks to the Blackfoot traders. In 1785 the NWC established Pine Fort near the old French Fort la Reine on the Assiniboine River. That post quickly became the NWC's base for trade with the Mandan and Hidatsa villages two hundred miles to the southwest. The flow of European goods to the Mandan and Hidatsa villages grew, and thus Crow access to these goods probably also increased.[20] The Crows were still badly outgunned by their enemies who traded at the Saskatchewan River posts. Even in the 1790s the Crows may have depended as much on the flow of guns northward from New Mexico as they did on guns coming west from the Missouri villages. When a combined Piegan and Shoshoni party attacked a party of Crows in 1793, they seized two Spanish guns.[21] The Spanish in New Mexico remained indifferent to the possibility of trade with Indians. Edward Umfreville saw Spanish horses, brands, and goods among the Blackfoot bands in the 1780s, but a Shoshoni slave woman among them told him that "it is not peltry that they [the Spanish] come principally in quest of."[22] In 1786 the Comanches in New Mexico, devastated by war and disease, became more peaceful. Peace encouraged Comancheros (New Mexican traders) to travel and trade in parts of the southern Great Plains and Great Basin, spreading Spanish goods as far as the northern plains.[23] To seal their peace, between 1786 and 1804 the Spanish distributed over 18,000 pesos' worth of European goods, including guns, to the Comanches. Some of these goods spread across the plains in trade and warfare,[24] although the northern Shoshoni bands must have acquired only very few of these guns.

Compared to the Blackfoot and Cree bands, the southern bands were very poorly armed in the 1780s and 1790s. Umfreville, who was on the Saskatchewan River from 1784 to 1788, saw Blackfoot shields of "drest leather, which are impenetrable to the force of arrows" but were useless against guns, which suggests that the Crows and Shoshonis rarely used firearms in warfare.[25] Even in 1793 Peter Fidler watched the Piegans practice war games with bow and arrows and shields, which may indicate that their adversaries still relied on those weapons. He met Kutenais equipped with stone arrowheads, hatchets made of deer horns, and wedges of stone.[26] Weaponry continued to make the decisive difference in warfare. David Thompson, after explaining that Indians were happy to give up fine pelts to get awls, needles, and European fire-making tools, added that "iron heads for their arrows are in great request but above all Guns and ammunition." This, he explained, was because "a war party reckons its chance of victory to depend more on the number of guns they have than on the number of men."[27]

The northern coalition capitalized on its military superiority. Any hiatus in warfare between the Piegans and Shoshonis after the smallpox epidemic of 1781 lasted only a few years, according to Young Man's account. He explained that a group of "Snakes" resumed the continual warfare two or three years after the epidemic by killing a small party of Piegans who had traveled to the Rocky Mountains to hunt.[28] Traders' records of the late 1780s and the 1790s reveal that plains Indians often went to war in summer and winter. As a result the Blackfoot bands, particularly the Piegans, dominated the northwestern plains at least as far south as the Oldman River basin.[29] By 1792 the Blackfoot bands were raiding well south of the Missouri River.[30] For over two months ending in January 1787 the Piegan warrior Kootenae Appe led an expedition to the south, finding no enemies "but the Black People (the name they give to the Spaniards [Utes]) from whom they had taken a great many horses and mules."[31] While Blackfoot bands probably rarely hunted in the Missouri drainage at this time, their warriors made the area north of the Missouri very dangerous for Shoshoni, Crow, Flathead, and Kutenai bands. The Blackfoot bands confidently raided their southern neighbors, augmenting their horse herds and dominating ever larger areas of the northwestern plains.

The northern bands also used diplomacy to complement their warfare. In June and November 1792 some Shoshonis and Kutenais sued Piegan bands for peace.[32] After some disagreement among themselves the Piegans agreed to discussions, at least in part because a truce gave them an opportunity to dazzle their enemies with their vast military superiority. In December 1792 three Shoshoni emissaries visited large Piegan encampments near the Highwood River. When one of them arrived at Sakatow's band, the Piegans ordered HBC traders visiting them at the time to wear their best clothes in order to impress the Shoshoni man.

Peter Fidler reported that "the Snake Indian viewed us from head to foot & from foot to head—with the greatest attention felt at our skin in places and expressed great astonishment at us—particularly at our having a different coloured hair from any Indian." When they sat down to smoke that sunny December day, Fidler seized an opportunity to endear himself to his hosts:

> I lighted it [a pipe] with a burning Glass that was fixed in the top of my Tobacco box—he eyed me all the time with the most circumspect attention but when he saw the pipe smoake—by means of the Glass—he jumped up & wished to be farther from me—as he thought I was something more than common, to light a pipe without fire—& the Indians we was with took good care not to let this good opportunity slip, to extol us in a very high manner to him & they told the poor fellow such unaccountable stories relating to our conjurations that was very ridiculous—but magnifying us in his Eyes.

The Piegans insisted that Fidler display his astronomical instruments and told their visitor that he used them to predict the future.[33] The Piegans also exhibited goods they acquired at trading posts and sent the Shoshonis off with some free samples.

At the end of their visit the Shoshoni men invited several Piegans to their camp about a fifteen-day walk to the south, well south of the Missouri River. After generously hosting the Piegan visitors for about six weeks, the Shoshonis gave them fine horses that carried them home in eight days.[34] That winter other Piegan and Blood men also visited the Shoshonis, who convinced them to join in attacks on Crow bands. For example, seventeen Blood men joined a Shoshoni attack that left thirty-two Crow men dead. During the same winter five Piegan men participated in a far less successful Shoshoni raid. Except for a Piegan man killed by a well-aimed Crow arrow, the aggressors suffered no injuries. In contrast to the prevailing patterns at the time, Blood, Piegan, and Shoshoni warriors waged war together against the Crows. Further complicating the picture, even as these Blood and Piegan men visited Shoshoni bands, other Blood and Piegan parties raided horses from other Shoshoni bands.[35]

The events of the winter of 1792–93, when some Blood and Piegan warriors joined Shoshoni raids on the Crows and some Piegans traded with Kutenai bands, illustrate how complex relationships among indigenous groups on the Great Plains were. We cannot adequately understand patterns of interaction among these bands unless we assume that raids and skirmishes frequently occurred among generally peaceful plains Indian bands just as episodes of peaceful trade and cooperation punctuated warfare between generally hostile bands.[36] Facile classifications of "allies" and "enemies" are inadequate in this regard. Relationships between bands, even bands of the same ethnic group, ranged along a continuum from close and persistent friendship to inveterate and continuous enmity. The impetus and ability to reconcile was usually greatest among related bands and among bands that relied on one another equally. A band that enjoyed clear military advantages over the other bands had greater freedom to take advantage of its neighbors, and the weaker bands were often compelled to tolerate abuses. Aggression did not necessarily sever relations, just as trade and military cooperation did not necessarily cement them. This fact is as important for understanding the relationship between the Blackfoot and the Shoshoni bands as it is for understanding the relationship between the Blackfoot and the Cree bands.

Trade and military cooperation clearly occurred among generally hostile bands. Conversely, hostile incidents were frequent on the northern plains even when groups lived in general amity. For many years patterns of symbiosis acted to ensure that occasional raids, even skirmishes involving deliberate killings, were tolerated and appeased among bands of the northern coalition. As the more powerful partners, Cree and Assiniboine bands had greater liberty to exploit and offend than the Gros Ventre and Blackfoot bands did, but the relationships usually exhibited goodwill.

Mutual vulnerability to and dread of the Cree, Assiniboine, Blackfoot, Gros Ventre, and Sarcee bands generally led the diverse southern bands to make common cause. Meanwhile, as a result of changing circumstances in the Saskatchewan basin, the powerful northern bands were increasingly at odds. The arrivals of European goods and of Euroamericans themselves are two distinct phenomena with very different implications for the people of the northern plains. The arrival and spread of European goods may have created but certainly strengthened interdependence among northern bands between 1730 and the 1770s. The arrival of

Euroamericans did the opposite.[37] The expansion of the provisions trade and the establishment of direct trade between Europeans and various plains bands undermined the mutually beneficial relationship between various Cree and Assiniboine bands and Blackfoot and Gros Ventre bands even as it enhanced the power of Cree and Assiniboine bands compared with the Blackfoot and Gros Ventre bands.

After the early 1780s the Blackfoot and Gros Ventre bands dealt directly with Euroamericans, rarely with Cree and Assiniboine traders. They had no need for Cree or Assiniboine military assistance against their weak enemies. Compared with the privileged position of the Crees and Assiniboines, the Blackfoot and Gros Ventre bands held a peripheral role in trade networks. For many goods, services, and furs Euroamerican traders relied on Cree and Assiniboine bands with whom they had already forged relationships. In contrast, "the *Gens du large* consisting of Blackfeet, Gros Ventres, Blood Indians, Piedgans &c. are treated with less liberality, their commodities being cheifly Horses, Wolves, Fat & Pounded meat which are not sought after with such eagerness as the Beaver."[38] The influx of wolf skins that accompanied the beginning of direct trade with the plains bands glutted the market, and wolf-skin prices in Europe fell in the 1780s. In 1789 the HBC dropped its standard for wolf furs from 2 MB per pelt to its older standard of 1 MB. Plains bands resented the change. Thomas Stayner noted of one Piegan band in 1790 that "these Indians are exceeding hard to deal with, the reducing the Standard of Wolves from two to one Beaver, will I fear injure this Trade much, as the Master has it not in his power that Encouragement, for those Skins, which was formerly given."[39]

Compared to the Crees and Assiniboines, the Blackfoot and Gros Ventre bands' position in the trade actually declined in the 1780s. If they had still been interdependent, they might have preserved cordial relations. But only their traditional friendships and kinship ties held them together. For the Blackfoot bands these ties were enough to overcome antagonisms for several more years, but relations did sour noticeably in the 1780s. Frequent horse raids were the main irritants.

Two factors influenced the distribution of horses on the Great Plains historically. The first factor, only important before the 1780s, was the rate of diffusion from the source in New Mexico. Between 1780 and 1800, however, the density of horse populations across the interior of North America roughly stabilized. Groups that were rich in horses by that time tended to remain rich in horses, while those poor in horses remained poor. After 1780 the distribution of horses had less to do with diffusion from the source than with environmental factors that limited the horse population. Horses became most plentiful on the southern plains, among the Kiowas, Apaches, Comanches, and Osages, and in the Columbian plateau, among the Interior Salishan groups, including the Flatheads.[40] Groups on the western plains were generally richer in horses than those on the eastern plains.

The severity and length of winters and the depth of snow appear to have been the most important factors in determining the density of horses across the plains and west of the Rocky Mountains.[41] Horses are far less well adapted to cold climates than are bison; thus the gradient in the density of horse populations was steep. On and near the northwestern plains the Flatheads, Shoshonis, and Crows consistently had the largest horse herds. Among the Blackfoot bands the Piegans had the largest horse herds and the Siksikas the smallest.[42] Other groups, such as the Gros Ventre and Sarcee bands, had still fewer horses. The Plains Crees and Assiniboines had fewer horses than almost every other group on the Great Plains.[43]

Maintaining domestic horses demanded considerable time, energy, and expertise.[44] The harsher the environment, the more onerous the demands of horse care for owners, especially when they considered the risk that despite their best efforts their horses might escape, be stolen, or succumb to severe weather. Horses were not pets; they were beasts of burden. Winter use was so hard on them that Cree and Assiniboine bands in areas of the northeastern plains often released their horse herds at the beginning of the winter and collected any survivors they could find in late winter.[45] They had little success in horse breeding and put little effort into horse husbandry.[46] In severe winters the Blackfoot often tried to rest weak horses by relying on dogs, but nomadic hunters sometimes could not give their horses the rest, food, shelter, and protection from predators that they needed to survive. Where winter kill was typically high, people found it more rewarding to raid their enemies for horses than to expend the energy needed to maintain and augment horse herds through careful husbandry and breeding.[47]

Traders in the eighteenth and nineteenth centuries repeatedly noted that the Plains Cree and Plains Assiniboine bands were the most notorious horse raiders on the northern plains. Unable to maintain their herds naturally, these groups had to replenish their horses from their neighbors' herds. As long as they acquired horses from the Blackfoot and Gros Ventre traders or got them from the abundant herds of their southern enemies in combined raids with the Blackfoot bands, their raids against the Blackfoot and Gros Ventre bands and the traders were relatively rare. Once the Blackfoot traders established regular direct trade with Euroamericans early in the 1780s, however, Cree and Assiniboine bands had little to exchange with the Blackfoot bands. They continued to join expeditions against the Shoshonis and Crows, but they also relied more heavily on horse raids against the Blackfoot and Gros Ventre bands and Euroamericans. Gradually and almost inevitably the relationships among these bands soured. Umfreville described the process in 1790: "Many broils and animosities among the natives, originate from a desire of being in possession of these animals. One party generally commences hostilities by stealing horses of their adversaries, and they in turn retaliate; so that at length a mutual resentment takes place, and war becomes absolutely necessary."[48]

Cree and Assiniboine raids on Euroamericans' horses began in the late 1780s when a series of bitter winters decimated the horse herds of the northwestern plains. Because they lived in settled establishments, committed considerable effort in gathering hay and sheltering horses, and had greater opportunity to rest weakened horses, the traders kept horse herds in environments where Indian bands could not. Trading posts became targets of repeated raids. Cree and Assiniboine bands, particularly those from the Swan River region of the northeastern plains (where horse mortality was the highest), traveled long distances to raid horses from the Indians and Euroamericans of the northwestern plains. In contrast, although horses were "their principal inducement in going to war," especially before the 1820s, the Blackfoot bands rarely raided the herds of traders or Cree and Assiniboine bands except in retaliation.[49] They directed their raids at the abundant horses of the Crows, Shoshonis, and Flatheads.

Acrimonious incidents in the Saskatchewan River region grew frequent between 1786 and 1788, putting the broad coalition of Blackfoot, Gros Ventre, Cree, and Assiniboine bands in jeopardy. In May 1786 traders learned that some Piegans had killed one Cree and wounded another near the Eagle Hills. In the summer of 1787 William Tomison grew concerned about the increasing tensions among the Indians near Manchester House. In July some Crees of the upper Saskatchewan River apparently attacked and killed some Bloods and Sarcees, forcing the Cree band to flee downstream to the South Branch House area for fear of retaliation. For the rest of the summer the arrival of Blood bands at Manchester House sparked anxiety, and HBC traders feared that Blood and Piegan bands would avoid Manchester House during the subsequent winter. In August a Piegan band, almost certainly Young Man's band, arrived. Manchester House journals reveal that "their chief business is to make it up between the blood Indians and Crees and Sussew Indians to be at peace and all to come to the Houses as before."[50] Young Man's connections with the Cree bands made him particularly well suited to mediate in this case. Other Blood and Piegan bands remained reluctant to visit Manchester House that winter. Growing tensions over the next year convinced Tomison to abandon Manchester House for the summer of 1788.[51]

There is no evidence, however, of further violence between the Blackfoot and Cree bands in the late 1780s and early 1790s. The relationship between them seems to have improved again. On 28 January 1792 William Tomison noted that a tent of Blood Indians was camping at Manchester House "to wait the arrival of some Nehethewea [Cree] Indians to know whether they are to keep peace with them or not."[52] During the winter of 1792–93 small Cree parties camped, often with Piegan bands, in the Red Deer valley and even south of the Bow Valley.[53] In the spring of 1795 Duncan McGillivray noted the arrival of a large party of Sarcees, Crees, Piegans, and Bloods at Fort George and complained that "the Crees are quite pitiful this Spring having amused themselves during the Winter with smoking & feasting along with the Piegans."[54] Did the relatively mild and snow-free winters of the early 1790s help ease relations between the Blackfoot bands and the Cree and Assiniboine bands by reducing the winter kill among horses?

Cree-Blackfoot relationships changed in the 1780s and early 1790s, but it is not clear what course the Sarcees were taking. References to the Crees fleeing Manchester House for fear of the Sarcees and Bloods in 1787 are juxtaposed with Umfreville's report of a Sarcee attack on a Blood band in the summer of 1787 or 1788. According to Umfreville, the Sarcees in the late 1780s maintained "a close alliance with the Nehethaway [Crees] rather to profit by their protection, than for any mutual esteem."[55] Perhaps the Sarcee bands were divided in their policies. In early 1788 they fought among themselves. Although the reasons for the fighting are unclear, at least four men were killed and several more injured. The conflict led some Sarcees to flee their kin and camp with some Cree bands.[56] Some Sarcees appear to have camped with Crees for some years thereafter.[57] Other Sarcees, however, were associated with Blackfoot groups in the 1790s. A Piegan-Sarcee party, for example, arrived at Fort George in 1794.[58] Easing Blackfoot-Cree relations in the 1790s probably allowed the Sarcees to maintain cordial relations with Blackfoot and Cree bands.

As relations among bands in the Saskatchewan River basin fluctuated, the behavior of plains bands at trading posts also changed. When indigenous communities grew more familiar with Euroamericans, their admiration for the newcomers quickly diminished.[59] Their policy toward traders, however, was also influenced by changing relationships among Indian bands. Although they tried to reconcile differences among bands, traders remained closely associated with Crees and Assiniboines. They had been familiar with these bands, who were the most dependable suppliers of furs and services, for many years. By the 1780s kinship, friendship, and trading ties between Cree communities and traders were well established. The Blackfoot and Gros Ventre bands never had the same access to European goods or confidences as did the Crees. When their relations with Cree bands were strained, and the military superiority of the Crees and Assiniboines consequently became more problematic, Blackfoot and Gros Ventre bands naturally resented the traders who supplied their rivals.

During the 1780s Blackfoot, Gros Ventre, and Sarcee trading parties became increasingly aggressive at trading posts.[60] Until the mid-1780s traders wrote glowing reports about the Piegans and Bloods, but this changed in the late 1780s. In the summer of 1789 Tomison left an unusually large number of traders at Manchester House because "the Indians are much more daring then they used to be."[61] That autumn Thomas Stayner described how a Siksika man, unsatisfied with trade terms, walked into the master's room at Manchester House with a gun. The confrontation ended peacefully, but it hinted at the increasing tensions. In the summer of 1791 Isaac Batt unwisely agreed to travel with a Blood band that was little known to the trader and had run away with women from another Blackfoot band. Batt was murdered for the horses, guns, and other articles he took with him. In 1793 a group of Bloods came and simply left because they were unsatisfied with the trade terms. The traders' opinion of the Sarcees also soured in these years. At Manchester House during the late 1780s the Sarcees began to earn a reputation as difficult traders.[62]

The traders' opinion of the Gros Ventres changed much more dramatically. The years between 1782 and 1794 are the most tragic era in Gros Ventre history. The period must have begun with great optimism. The Gros Ventres were able to buy more European goods at lower prices than ever. They enjoyed generally cordial relations with their immediate neighbors and huge military superiority over their enemies. Fifteen years later their situation was radically different. Tragically, the devastating smallpox epidemic of 1782 was only the dramatic prelude to decades of difficulty.

Strong friendships tied Gros Ventre and Blackfoot communities together before and after 1780. In the 1770s and 1780s, however, they took different paths as they faced unique opportunities and challenges. Located east of the Blackfoot bands, Gros Ventre bands felt the consequences of Euroamerican expansion earlier and more intensely; unlike the Blackfoot bands, they had kin a considerable distance to the south. The documentary evidence for Gros Ventre–Arapaho interaction dates from the first years of the nineteenth century, but there can be little doubt that interaction occurred long before that, perhaps continually since their separation. The Gros Ventre–Arapaho relationship meant that the Gros Ventres were much more oriented toward and familiar with the region south of the Missouri than were the Blackfoot bands. Maps drawn by Siksika and Gros Ventre leaders in 1801 and 1802 make this clear. On a map drawn by Old Swan in 1801 landforms south of the Missouri River are identified by their Gros Ventre names rather than their Blackfoot names (see figures 7.2 and 7.3). A map drawn by a Gros Ventre informant depicts a region extending as far as New Mexico, much farther south than any of the Siksika maps (see figures 7.4 and 7.5). It shows the location of several Arapaho–Gros Ventre bands well south of the northwestern plains. The opportunities and challenges the Gros Ventre bands faced affected their behavior and prospects significantly in the 1780s and 1790s.[63]

The Gros Ventre bands were more peripheral to Euroamerican trading networks than the Blackfoot bands were. Because they could supply few of the furs that Euroamericans valued most highly, and because the Gros Ventres were a small linguistic group, traders did not send emissaries to them and never attempted to learn their language.[64] Compared to other

		Tents
1	Choque – Mud Houses Indians	150
2	To sap poo – Crow Mountain	200
3	Ama cops six sue – Sessews Souk	70
4	Sip pe tah ke – Wrinkled	90
5	Kix tah ka tap pee – Beaver	50
6	Choque – go to war with No 1	160
7	Nee coo chis ak ka – Tattood	80
8	Sin ne po tup pe – Grey fox	40
9	Ke ta kap sum – Garter	30
10	To hoo is too ye – Hairy or Beard	50
11	Ak kin nix sa tup pe	40
12	Pick et a tup pe – Rib	100
13	Oo aps six sa tup pe – Thigh	20
14	Oe sa tup pe – scabby	100
15	Mae que a tup pe – Wolf	200
16	Mut tah yo que – Grass Tents	180
17	Mem me ow you – Fish Eaters	20

		Tents
18	Ne chick a pah soy a Particular Root	60
19	Mis che tap pe – Woody	100
20	Six six chicks sin na tap pe Sessew Snak 2d	50
21	Poo can nam a tup pe – Pearl shell	70
22	Six too ki tup pe – Black	200
23	Cut tux pee too sin – Flatt heads	50
24	Cum men na tup pe – Blue mud	60
25	Ap pa tup pe – Varmin or white	80
26	To ke pee tup pee, those that collect shells	
27	Oe cook sa tap pe – Pabling	30
28	At cha tap pee – Snare Ind	18
29	Cut tux in nah mi – Weak Bow	18
30	Patch now	10
31	Catton na	22
32	Pun nus pee tup pin – Long Hair	100

M L K I N H G

12 13 14 15

7 8 9 10

3 2 4 5

Attee nut chay or Sheep river

Ewow se nut chay or Red Deers river heavy rapids, deep & very strong current

Bay in nut chay or Pearl river

Issee sic ah cha or Snake mountains

Sun nee kaw sis or warm water river

Nat che keck or Snowy mountain Rocks

Bas caw sis or Cape river

Sic see ah Rocks Snake mounts

Ki apho ta he

From B to C	2 Days
C to D	5
D to F	3
F to G	3
G to H	1
H to N	1
N to I	2
I to K	1
K to L	5
L to M	9
D to E	1

6

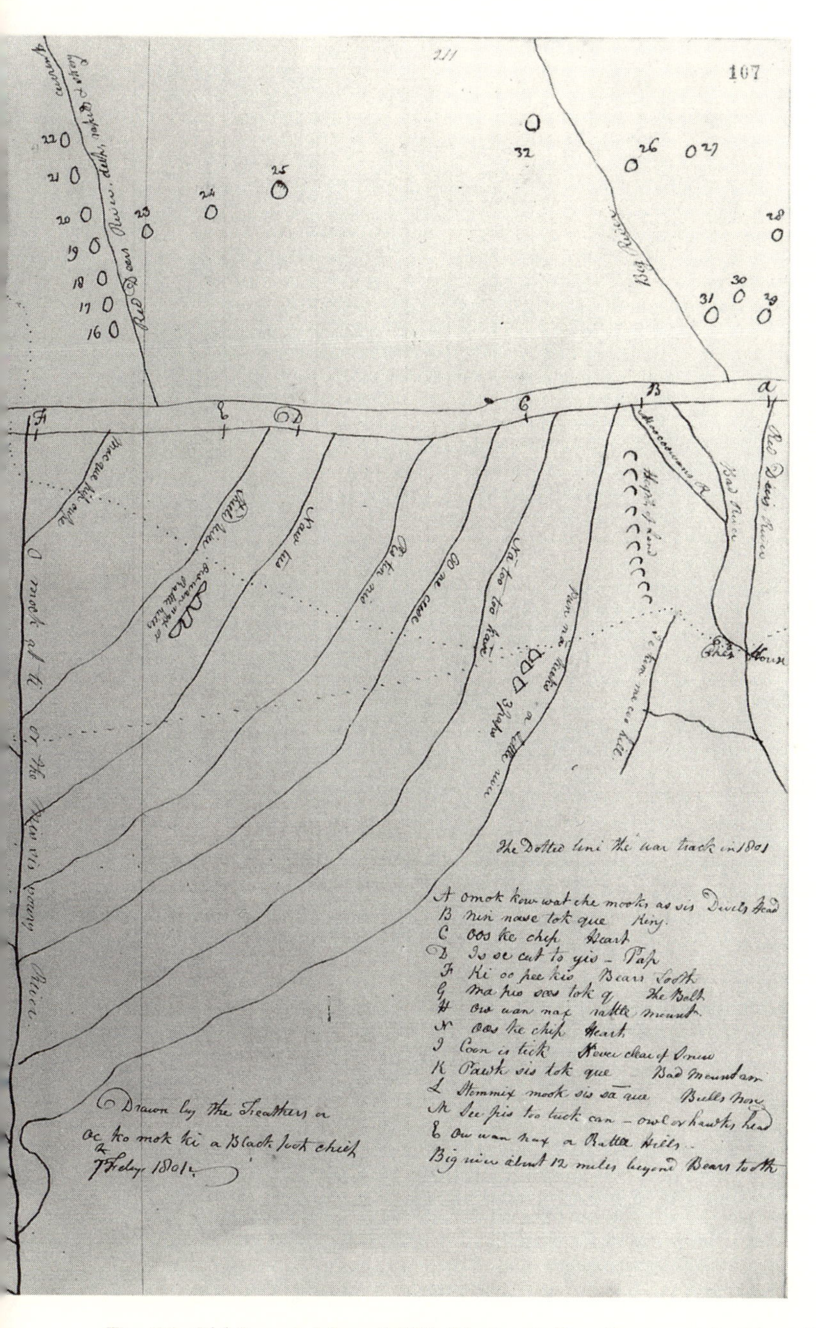

Fig. 7.2. Old Swan's Map of 1801. Hudson's Bay Company Archives, Provincial Archives of Manitoba, E.3/2, fols. 106d–7.

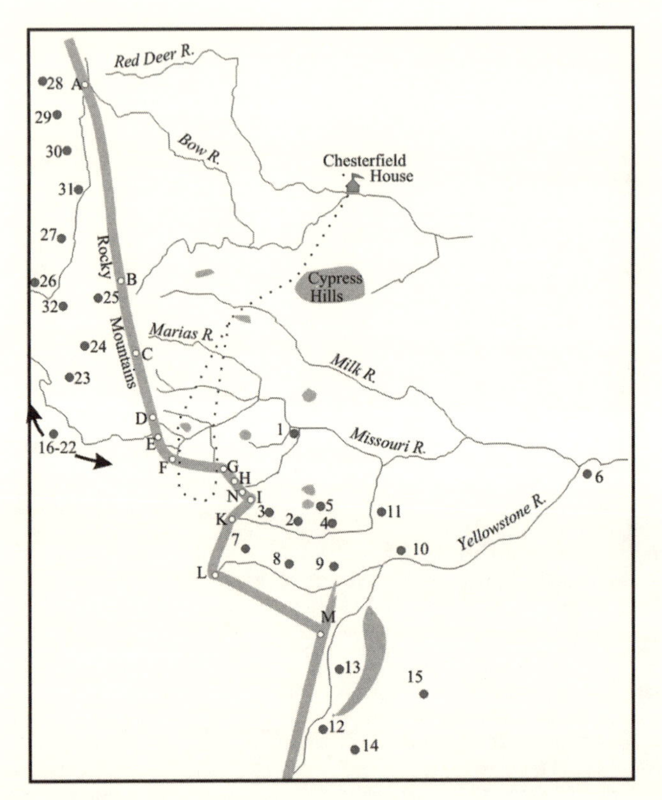

Fig. 7.3. Translation of Old Swan's Map of 1801. Old Swan's map depicts the northwestern plains at the height of Blackfoot power in the region. The map, which indicates the approximate locations of bands not associated with the northern coalition, shows that the enemies of the Blackfoot bands were restricted to the margins of the region. Legend: A. Devil's head; B. Chief Mountain; C. Heart Butte; D. Haystack Butte; E. Birdtail Buttes; F. Beartooth Mountain; G. Belt Butte; H. Wolf Butte; I. Point in the Big Snowy Mountains; K. Coffin Butte; L. Canyon Mountain; M. unidentified point in the Pryor Mountains. Bands: 1. Crow or Hidatsa; 2. Crow; 3. Crow; 4. Cheyenne; 5. unidentified; 6. Crow or Hidatsa; 7. Arapaho; 8. perhaps Arapaho or Cheyenne; 9. perhaps Arapaho or Cheyenne; 10. perhaps Cheyenne; 11. perhaps Cheyenne or Arapaho; 12. unidentified; 13. perhaps Cheyenne or Arapaho; 14. Northern Cheyenne; 15. Pawnee; 16. Shoshoni; 17. unidentified; 18. unidentified; 19. unidentified; 20. Shoshoni; 21. unidentified; 22. Ute; 23. Flathead; 24. Nez Perce; 25. unidentified; 26. Pend d'Oreilles or Kalispel; 27. Pend d'Oreilles or Kalispel; 28. Snare; 29. Lower Kutenai; 30. unidentified; 31. Upper Kutenai; 32. unidentified. Travel time (in days) between locations: B to C, 2; C to D, 5; D to F, 3; F to G, 3; G to H, 1; H to N, 1; N to I, 2; I to K, 1; K to L, 5; L to M, 9; D to E, 1. Adapted from Binnema, "Indian Maps"; and Moodie and Kaye, "The Ac Ko Mok Ki Map," 4–15.

northern bands, the Gros Ventres could purchase relatively few European goods. Located outside the chinook belt and near powerful Cree and Assiniboine bands, Gros Ventre horse herds suffered more in winter and were more subject to raids than were Blackfoot herds. The Gros Ventre bands were more dependent on wolves than the Blackfoot bands were and felt the effects of the HBC reduction in the trade standard for wolf furs more acutely. As the Gros Ventre bands' ability to secure European goods diminished and their relationship with their well-armed Cree and Assiniboine neighbors deteriorated, their military position began increasingly to resemble that of the Shoshonis and Crows. In a desperate effort to defend themselves, the Gros Ventre bands attacked Euroamerican trade posts in 1793 and 1794. Their efforts only incited greater hostility, and the 1790s witnessed the beginnings of the gradual withdrawal of the Gros Ventres toward the south.

The reduction in the HBC standard for wolf furs in 1789 could not have come at a worse time for the Gros Ventres, for it reinforced the impression that the Euroamerican traders were firm allies of the Cree and Assiniboine bands. By the late 1780s relationships between local Cree and Assiniboine bands and Gros Ventre bands were unfriendly. The Cree and Assiniboine bands were much better armed, because they had greater access to posts (having largely chosen their locations) and to a wider range of valuable fur-bearing animals and could also profit by provisioning the traders.

While the Gros Ventres clearly benefited from their access to Euroamerican goods, they were at a disadvantage when compared to the Crees and Assiniboines. This would not have been a serious problem if relations had remained friendly, but they had not. Cree bands that adopted an equestrian plains lifestyle in the 1770s and 1780s frequently raided Gros Ventre horses, producing serious conflict. The basis for friendship—mutually beneficial trading relations and dangerous mutual enemies—had disappeared by the 1780s. Relations between the Gros Ventre and Cree and Assiniboine bands were very tense by 1787, but they seemed irreparable after 1788.

During the unusually cool spring of 1788, after a cold, snowy winter, a small band of Gros Ventres nearing posts in the Eagle Hills was attacked by a band of about ten tents of Crees from the lower South Saskatchewan River area, who "fell upon them and killed the leading man, after which they cut off his arms, head, Private Parts and took out his bowels and then

Chetow

Chetow

		Tents	
1	Cock kaw etch – Flatt Heads	50	
2	Cock kaw yaw etch Do	40	
3	See see an nen Snake Inds	130	
4	Ow win nen Con mount	100	
5	Ee thā chee nā	30	
6	See see an nen	10	
7	May aytch at chot	20	
8	Nan ni en	100	Lodg
9	E tā seen	100	
10	Sot tan	30	
11	Neet chay in in	100	
12	Chow win in	20	
13	E chaugh ā nen	10	
14	On now win	10	Lodg
15	Now wā ben nen nach	10	
16	Wā tan nitch	200	
17	Now wā se se an nen	300	
18	See see an nen ne hand thee – Spanish Settlements		
19	Oth thay in in	100	
20	She ne nen	100	
21	Now watch e ni in	20	Lodg
22	May aytch e chot	10	
23	Chow win nen	8	
24	Shot thok ki in in	13	
25	Beth thow in in	20	
26	Ben cot chaw hatch	40	
27	Wan nuk ki an	40	
28	a. k. thi a wootch	20	
29	ā heth thoo	15	
30	Been nen in	100	

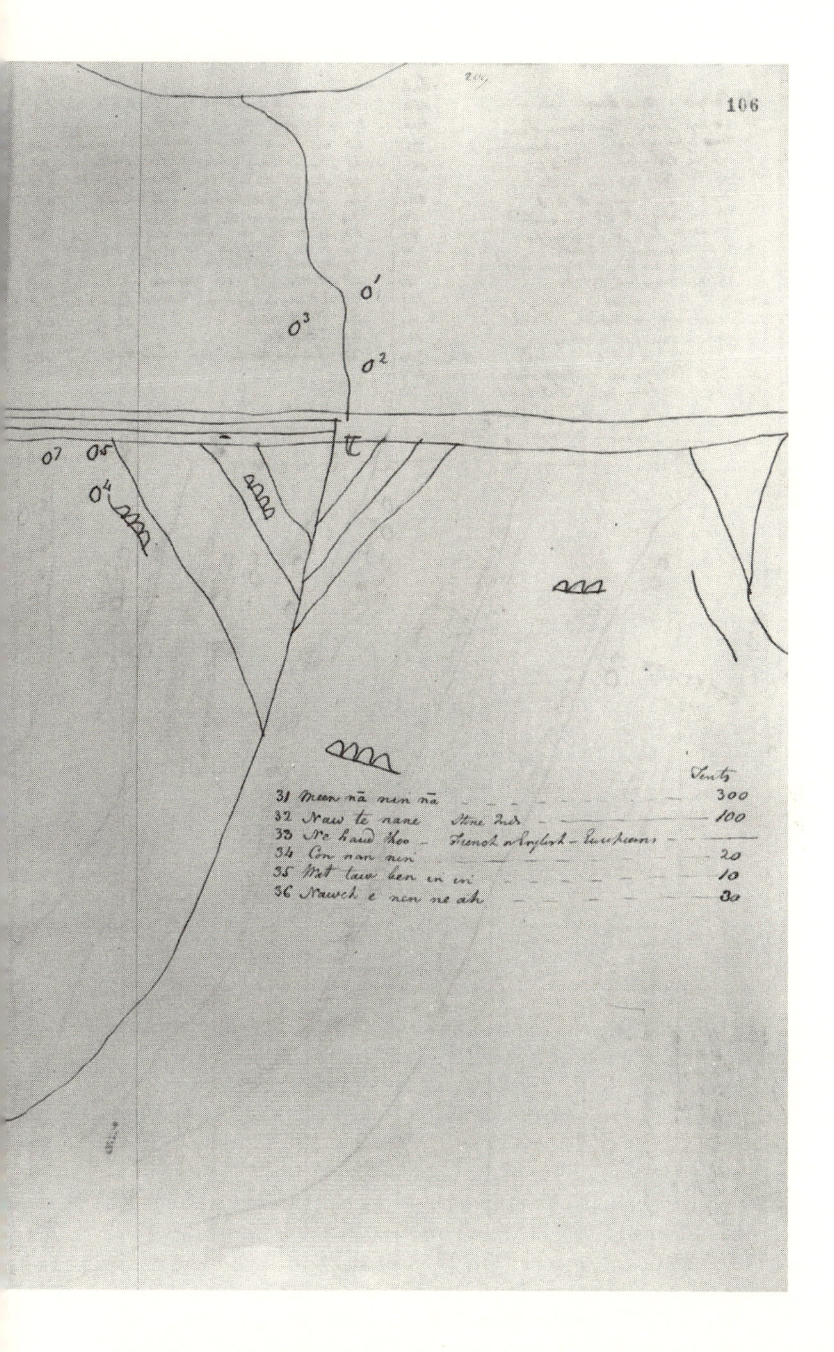

Fig. 7.4. Map by Unidentified Gros Ventre Cartographer, 1801. Hudson's Bay Company Archives, Provincial Archives of Manitoba, E.3/2, fols. 105d–6.

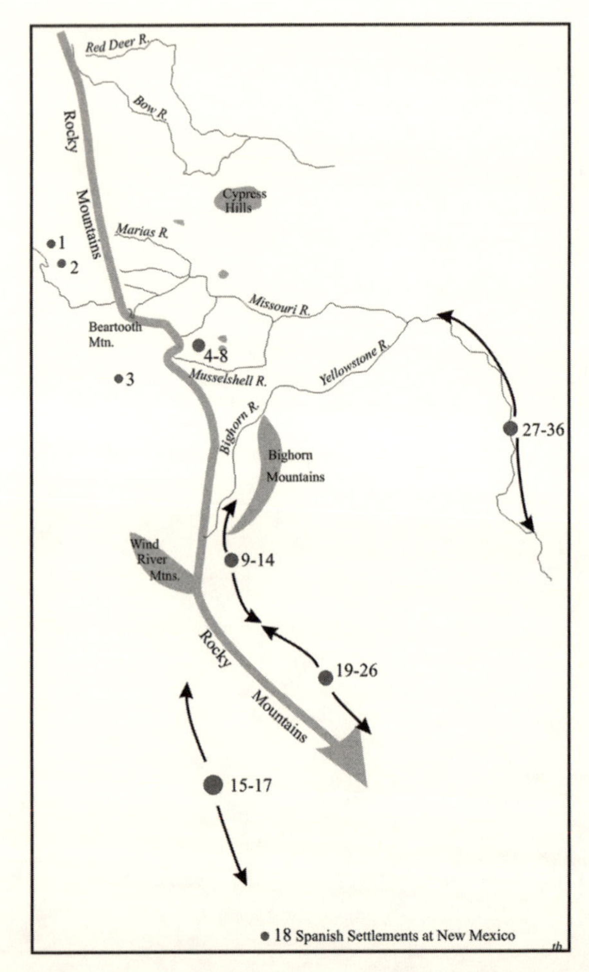

Fig. 7.5. Translation of Map by Gros Ventre Cartographer, 1801. Legend (numbers correspond to the number of tents occupied by each group): 1. Flathead (50); 2. Flathead (40); 3. Shoshoni (130); 4. Crow (100); 5. Cheyenne (30); 6. Shoshoni (10); 7. unidentified band (20); 8. Arapaho (kin of the Gros Ventres) (100); 9. Cheyenne (100); 10. unidentified (30); 11. Kiowa (100); 12. Crow? (20); 13. unidentified (perhaps Crow) (10); 14. Arapaho (kin of the Gros Ventres) (10); 15. unidentified (10); 16. Ute (200); 17. Shoshoni (300); 18. Spanish settlements at New Mexico; 19. Comanche (100); 20. Pawnee (100); 21. Southern Arapaho (20); 22. unidentified (10); 23. Crow? (8); 24. Kiowa Apache (13); 25. Arapaho (20); 26. unidentified (perhaps Arapaho((40); 27. Hidatsa (40); 28. unidentified (20); 29. unidentified (15); 30. unidentified (100); 31. unidentified (300); 32. Assiniboine or Sioux (100); 33. French or English Europeans; 34. Arikara (20); 35. unidentified (10); 36. Sioux or Assiniboine (30). Adapted from Binnema, "Indian Maps"; and Gibson, Untitled manuscript (HBCA PP 1988-14).

took what furrs they had untraded from them," which the Crees then "traded with [the?] impudent villains of Canada."[65] The troubles prompted the HBC to abandon Manchester House for the summer of 1788. Because of the tensions, even at South Branch House William Walker found it difficult to keep his Cree hunter in the area during the summer. That summer more "great disturbances" including an incident that left five Gros Ventres and one Cree dead led the Crees of the region to flee downriver and into the forests and the Gros Ventres to flee to the south, probably to visit their Arapaho kin.[66] The Gros Ventres did not return to the posts for two years.

The severe winters of the late 1780s aggravated the situation. The winter of 1787–88 was difficult, and a major snowfall on 23 October 1788 at the Manchester House area heralded the start of another bitter winter. By the end of November unusually deep snow made horse travel impossible on much of the northwestern plains. By late January 1789 "Indians from different quarters" told William Tomison that "the Snow is so very deep which has starved the greatest part of their horses to death, and what is alive cannot travel."[67] In January and February Sarcees, Crees, Bloods, and Siksikas complained that the severe winter had decimated their horse herds. On 16 May Tomison noted that "it has been one of the coldest Springs I ever knew, not one night as yet without a strong frost."[68] The weather took its toll on the horses and the bison. Traveling near the forks of the Saskatchewan that spring, Tomison saw "many grizzled, brown, and black Bears today feeding along the Shores on drownded Buffalo, of which their is great Numbers." The winter of 1789–90 was milder, but spring was even later. The North Saskatchewan River at Manchester House did not break up until 11 May 1790.[69] The string of hard winters and late springs must have depleted the horse herds of all the bands in the Saskatchewan basin by 1790, but especially those of the Crees and Assiniboines. They were probably very tempted to raid their neighbors' herds.

If hostilities disrupted the lives of Cree and Assiniboine bands in the Saskatchewan River forks region, they affected the lives of the Gros Ventres to a far greater degree. After the warfare of 1788 the Gros Ventres avoided the Saskatchewan River region. On 12 May 1790 William Walker finally noted the arrival of "a Great Tribe" of Gros Ventres with a large trade in furs. He explained that "these Indians has been off from any of these Settlements these 2 Years through some Differences that happened between them and the Southward Indians."[70]

In contrast to previous winters, the first three winters of the 1790s were warm and snowless, thanks to one of the strongest El Niño events ever known.[71] Perhaps the mild winters ameliorated the situation for the hard-pressed Gros Ventres, for the tenor of HBC journals suggests that tensions at the posts abated. In 1792 the NWC and HBC built establishments near the Moose Hills. The new Buckingham House / Fort George posts served local Cree and Assiniboine bands and became the favorite site for the Piegans, Bloods, Sarcees, and some Siksikas. Local Cree and Assiniboine bands, and some Siksikas and the Gros Ventres, continued to trade at Manchester House / Pine Island House. The Gros Ventres visited Manchester House often in the winter of 1792–93. But war broke out with renewed intensity in the summer of 1793. A large group of Crees from the lower Saskatchewan River and Swan River regions, together with some Assiniboines, discovered a small band of about sixteen lodges of Gros Ventres camped near the South Saskatchewan River. They attacked, killing 2 old men and 150 women and children, then fled and traded at posts in the Swan River region for fear of retaliation, not returning to the Saskatchewan until the following spring.[72]

The 1793 battle was a dramatic turning point in Cree–Gros Ventre relations. Although there had been frequent conflicts before then, this was the first time the Crees killed many of the Gros Ventres. The Gros Ventres faced a serious predicament. They could not ignore this aggression; but as long as the Crees enjoyed such military superiority, the Gros Ventres could attempt to negotiate a peace agreement only from a position of terrible weakness. A retaliatory attack would be suicidal. Not only were the Gros Ventres outgunned, but after any incident Cree bands easily retreated into the forests, with which the Gros Ventres were unfamiliar.[73] The Gros Ventres could withdraw permanently to the south, where they had kin and friends, as they had done temporarily a few years earlier. But withdrawal meant cutting themselves off from Euroamerican traders. The Gros Ventres had to debate a solution.

The traders were well aware of the rising tensions, but were naively blind to the danger they now faced. It was only in hindsight that Duncan McGillivray explained that "the Gros Ventres being intimidated from attempting any speedy revenge upon the Crees, formed the design of attacking us, whom they considered as the allies of their enemies."[74] The Gros Ventres, and some of their Siksika neighbors, evidently formed this design gradually. During the summer and autumn of 1793 Siksika bands at Buckingham House dealt very aggressively with the traders. Tensions rose during the unusually cold and snowy October of 1793 when a Cree man killed a Siksika man at Buckingham House after a quarrel, forcing the Siksika band to flee.[75] Recognizing the danger, HBC and NWC traders had agreed to abandon their South Branch Houses for the winter of 1793–94 until William Tomison let his bitter hatred of his NWC rivals cloud his judgment. He scuttled the deal.[76] Only with great difficulty could James Bird at South Branch House convince his Cree hunter to hunt for him that winter.[77] When a combined Gros Ventre and Blackfoot party pillaged Manchester House in October, Tomison and other traders were caught unprepared.

The Gros Ventres and Siksikas attacked Manchester House and Pine Island House to seize as many European goods as they could carry and to destroy what they could not, without injuring the traders themselves. The visit began normally. Several men came to announce that their combined Gros Ventre and Siksika party would soon arrive to trade the furs from their summer hunts. They returned to their families after the traders gave them the usual gifts of tobacco. Three days later, between 5 and 15 October, the band of about forty men arrived on the south side of the river. The traders ferried them over on boats. If the Gros Ventres had planned to kill traders, they let their best chance pass. They began the normal routine of trading their furs at the posts. Only then did they detect that only two traders were present at the HBC's Manchester House and spontaneously began pillaging the post and roughing up the men. They moved on the NWC post and apparently occupied most of it and began plundering until the NWC men took to arms. The attackers fled with some trade goods, about twenty horses, and two women from the post. Repulsed, the attackers returned to the less fortified HBC post

> and threaten'd the two men that if they did not immediately leave the House they would shoot them, they ransacted the house in every part, and carried away what they were in need of. All the Liquor they carri'd to the outer Gates & destroy'd it, and did not take or drink a drop of it, the loss is estimat'd at 3000 MBeaver belong to the Company besides the Men left all their own property. They went again to the Canadians house robb'd the Men of every thing they

> had—the Masters apartments being defend'd by their Men prevented their further progress, just as they had plunder'd the Mens house our People from up the River return'd in the Boats, which the Indians immediately secured and made a precipitate retreat across the River, where a Canadian fired and killed one of them.

None of the traders was killed, but evidently several other wounded attackers later died.[78]

At South Branch House James Bird expressed his "astonishment on hearing of a House being plundered by a People, I thought, the most rational and inoffensive in this part of the Country." Bird's surprise may seem unwarranted, but it reflects the traditionally amicable relations between the HBC and all the indigenous bands of the region, including the Gros Ventres. Until 1793 traders remained remarkably indifferent to the warfare that flared up around them. Even after the attack on Manchester House, however, Tomison understood that the traders themselves were not the Gros Ventres' main targets: "these always having been a peaceable People till now and what they have done I judge to be out of spite as they could not be revenged on the Southd and stone Indians for murdering so many of them last Summer."[79]

After the October 1793 affair the Gros Ventres and Siksikas who had plundered Manchester House grew more menacing. During the winter of 1793–94 they nursed their anger at the death of their kin. In early January a large band of about 150 Siksika men led by Big Man (O mok apee) arrived to trade at the Buckingham House and Fort George. According to Tomison, many of them "were very ill behaved several of them were cloathed in our Cloth and had a great many new Guns with them these I judge to have been conjunct with the fall Indians in the robbing of Manchester House last fall."[80] Duncan McGillivray believed that the band had returned to the posts to avenge the autumn deaths but found the forts too well guarded. Instead they stole about sixty horses, "threatening at the same time to *Scalp* the people & plunder the goods, as a Sacrifice to *appease the Spirits* of their deceased relations."[81] When they left the post they intercepted McGillivray and two other NWC men on their way back to the post after trading at another camp. They pillaged the traders' goods and "stript [them] to their shirts and had it not been near the House they must have perished."[82]

On 18 January 1794 a group of Siksikas and Bloods arrived at Buckingham House, but Tomison believed that they "behaved very rudely so that I was obliged to disarm them before they came into the House one of those that went to the Canadian House fired upon the Men that was waiting their arrival without any Provocation whatever; such behaviour I cannot account for." In February Tomison sent men from Buckingham House to retrieve what they could from Manchester House. There they met a large group of about 150 Siksikas and Gros Ventres who traded 1,200 wolf skins and some horses. The Indians stole fifty-two horses that night. According to Tomison, "had they not seen 50 armed men at the Canadian House its not known what lengths they would have gone to."[83]

Until the spring of 1794 the Gros Ventres had spared the traders' lives, but their reluctance to shed blood had evaporated by June. On 24 June 1794 they targeted South Branch House; if they had sought nothing but plunder in 1793, they now also planned to kill as many Euroamericans as possible and to destroy their trading facilities.

Everyone knew that trading posts were most vulnerable in the summer. In spring canoes embarked downstream for York Factory, leaving only a few people at each post. In 1794 the HBC left five men and five or six women and several children at South Branch House for the summer. On 24 June only two men, William Fea and Cornelius Van Driel, were at the post. James Gaddy Jr. was out gathering birch bark about fifty miles from the post, and two other men, Magnus Annel and Hugh Brough, in company with a Cree man known as "the Flute," were looking for the company's horses two or three miles from the post.

> They heard a great noise resembling the galloping of horses which hastily approached them, but judging them to be Stone Indians, as they knew them to be their Friends were not alarmed, the Southd Indian suspecting them to be Fall Indians, begged of them to go with him into a hammock that was just by while they passed but they were both deaf to his entreats, and paid no regard to them, however, the Southd Indn hasten'd into the thicket and hid himself, soon after to the amount of about one hundred Fall Inds made their appearance upon horse back, and riding up to our two Men alighted, kill'd and scalped them which they took away. They then proceeded towards the House.[84]

The neighboring NWC and HBC South Branch Houses were built on the south side of the South Saskatchewan River. The HBC post defenses were designed to prevent theft and mischief, not to withstand a concerted attack. According to Cornelius Van Driel, the HBC post was "surrounded with stockades that the 1st gale of wind we expected would level with the ground."[85] As the attack began, NWC interpreter Jacques Raphael was out on a recreational horse ride when he saw the large group of 100 to 250 (accounts vary) Gros Ventre and Siksika attackers.[86] Perceiving their hostile intentions, he raced back to the NWC fort in time for the four or five Canadian and three to five Cree men in residence to secure it. The two HBC men also shut the gates to their post before the Gros Ventres arrived and surrounded it. Obviously outnumbered but hoping to defuse the situation, the traders held their fire; but the attackers set the stockades on fire. Seeing William Fea through the cracks of the stockades,

> they fired at him and broke his arm—he immediately went through the Window at the back of the house into the Garden followed by Mr Vandriel the former laying down in an Old Cellar and the Latter in another one about 10 Yds apart, which very fortunately was full of rubbish which he covered himself with, presently the Inds broke open the Gates and entered the House and traced Willm Fea (by his blood) into the Cellar and Shot him dead. Mr Vandriel expecting every moment to Share the same fate, finding no more Men about the House they plunder'd it of every thing set fire to it and reduced it to Ashes, which loss is estimated at 4000 Beaver on their quiting the House they Stabbed Mags Annels Wife kill'd two of his Children, which they put onto their mothers Belly, three young women belongg to the Men that went to the Factory they took prisoner with them after plundering and destroying our house they went to the Canadians about 300 Yards distant intending to serve them in the same horrid manner.[87]

In contrast to the HBC post, the NWC's South Branch House was a well-built fort complete with block houses "about 40 ft long & wide and supported by four posts 20 f high with four or six small portholes on each, the logs &c 10 Inhs thick Proof against the Musquet ball." Louis Chastelain, four Canadians, and three Cree men defended the fort. They were so well protected that they killed five and wounded nine of the attackers without suffering an injury themselves. The attackers shot at the fort from behind a small rise for about half an hour, but

> when their War Chief L'Homme de Callumet a brave and undaunted Indian disparing of success from the mode of attack, which did not agree with his fiery nature, advanced a second time towards the Gates encouraging his Warriors to follow him; but he was interupted in the midst of his harrangue by a Shot from the Before mentioned interpreter which Streched him breathless on the ground, and the miscreants after recovering his body, retreated with mournfull lamentations for loss of their leader and threatening vengeance against the authors of his death.[88]

The attack on South Branch House changed life in the region dramatically. Traders quickly abandoned or fortified their more vulnerable positions while indigenous bands scattered. Within days Louis Chastelain abandoned his physically unscathed NWC post at South Branch House and established a post at Nipawin (Fort St. Louis). Next to this post James Bird replaced the destroyed HBC post with Nippoewin House. More permanent replacements for the South Branch Houses were built farther up the river in 1795 or 1796 at Peonan Creek, just below the forks of the Saskatchewan. The HBC named its post Carlton House; although it was safely within Cree territory, George Sutherland ordered James Bird to fortify it "in such a manner that no tribe of Indians dare approach it in a hostile manner in the future—the Muskettoons are well calculated to awe the Natives."[89] The NWC abandoned Pine Island House in 1794, and the HBC never rebuilt its sacked Manchester House. The HBC employees at Buckingham House abandoned their post and moved in with the NWC at Fort George until their post could be fortified. The Cree bands also withdrew from the Buckingham House region. In the fall of 1794, when it was time for the canoes to journey back up the river to the various posts, the HBC and NWC men traveled together for their mutual protection. The HBC sent only nine canoes to the upper posts, compared with sixteen the year before. The NWC Saskatchewan River outfit consisted of sixteen canoes, three fewer than the previous year.[90]

Although the attacks instilled great fear, the Gros Ventres launched them from a position of weakness, not strength and confidence. They represented a desperate attempt by the Gros Ventres to improve their position on the northwestern plains—and they were unsuccessful. Traders learned that the attackers immediately fled south, hoping to make common cause with the Shoshonis and "abandon this quarter for ever."[91] None of the Gros Ventres did abandon the region permanently in 1794, however. Soon the belligerents returned to be reconciled to the traders. Unable to avenge the killings, and unable to conduct normal business without peace, the traders quickly reconciled with the Gros Ventre and Blackfoot aggressors. They settled for preserving the memory of the events of June 1794 in their collective memory for decades. The Gros Ventres were never reconciled with the Cree and Assiniboine bands of the region. The hard winter of 1794–95 only encouraged the Crees to raid the Gros Ventres again.[92] The Gros Ventres' precarious position in the Saskatchewan basin deteriorated further after 1794, and they began to withdraw south toward the Missouri basin and their Arapaho kin.

CHAPTER EIGHT

The Apogee of the Northern Coalition, 1794–1806

From what they [the Siksikas] relate respecting the Gros Ventres *we have lost all hopes of seeing them this Spring. They are in the same dispositions as they were represented by our last accounts; that is, desirous of renewing their intercourse with us and our allies, but fearfull of a bad reception: for as they are naturally treacherous and vindictive themselves, it is reasonable to suppose that they suspect others of the same sentiments especially as they are conscious of having merited chastisement for their late depredations; and untill they can be assured that no violence is intended against them, at least as far as regards their persons I am affraid they will not be persuaded to visit us.*

—DUNCAN MCGILLIVRAY
1795

On the north side of the Missouri, near the Rocky mountains, resides a nation of Indians, who are numerous and who are the inveterate enemies of the Gros Ventres [Hidatsas] and the Crow Indians, and frequently fall on their hunting parties. They are called the Blackfoot Indians.

—CHARLES LE RAYE
1802

They [the Shoshonis] told me that to avoid their enemies who are eternally harrassing them that they were obliged to remain in the interior of these [Rocky] mountains at least two thirds of the year where the[y] suffered as we then saw great heardships for the want of food sometimes living for weeks without meat and only a little fish roots and berries. but this added Cameahwait, with his ferce eyes and lank jaws grown meager for the want of food, would not be the case if we had guns, we could then live in the country of buffaloe and eat as our enimies do and not be compelled to hide ourselves in these mountains and live on roots and berries as the bear do.

—MERIWETHER LEWIS
1805

Between 1794 and 1806 the Gros Ventres had few friends on the northern plains. Cree and Assiniboine bands attacked them when they could, and traders disliked them. Some of the Blackfoot bands were at peace with them but were preoccupied with their own fragile relations with Cree and Assiniboine bands. Among the Blackfoot and the Cree and Assiniboine bands, skilled diplomats defused hostile incidents and maintained a general state of peace. The northern coalition, minus the Gros Ventre bands, continued to hold together. The turn of the nineteenth century witnessed the Blackfoot bands at their peak of power and dominance on the northwestern plains. To the south the diverse bands of the southern coalition, now dominated by the Crows, resolutely hunted bison on the periphery of the northwestern plains, at least seasonally. Some bands even raided their powerful Blackfoot enemies, but they themselves could find no place where they were safe from the formidable "Raiders on the Northwestern Plains."[1] Their supply of European goods, especially from the middle Missouri villages, became more dependable, but their access to European weaponry remained inadequate to curb Blackfoot aggression. Meanwhile Euroamerican traders competed among themselves for the resources of the region, and their struggle continued to influence patterns of trade, warfare, and diplomacy.

Despite the increasingly volatile nature of life on the northwestern plains, Euroamericans continued to expand their trading operations across the northwestern interior of North America (see figure 8.1). Although the North Saskatchewan River basin had far fewer valuable furs than the Athabasca, traders still competed fiercely there. Having amalgamated

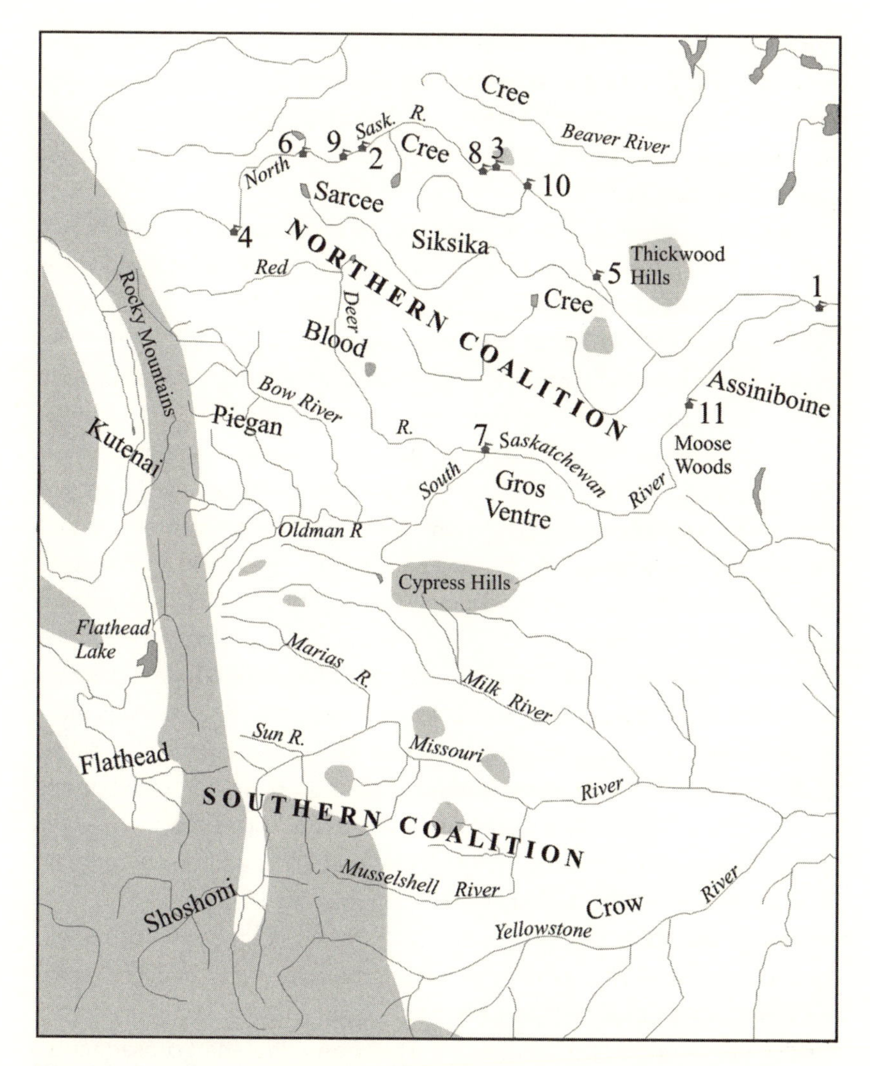

Fig. 8.1. Location of Euroamerican Trading Centers and Indian Groups, 1795–1806. Legend: 1. Carlton House I, 1795–1805; 2. Edmonton House/Fort Augustus I, 1795–1802; 3. Buckingham House/Fort George, 1792–1801; 4. Rocky Mountain House/Acton House, 1799–; 5. Somerset House, 1799–1800; 6. Nelson House, 1799–1801; 7. Chesterfield House, 1800–1802, 1804–5; 8. Island House/Fort de l'Isle, 1801–2; 9. Edmonton House/Fort Augustus II, 1802–10; 10. Fort Vermilion/Paint River House, ca. 1802; 11. South Branch House/Carlton House II, 1804–.

with, absorbed, or crushed its main competitors, the exceedingly prosperous NWC faced only feeble challengers from Canada between 1787 and 1795. Disaffected Nor'Westers David and Peter Grant organized weak opposition in 1793, but the NWC absorbed this short-lived and ill-fated partnership in 1796.[2]

The NWC, however, was about to face far stiffer competition. After the 1794 Jay Treaty, which stipulated that the British would abandon posts in the United States by 1796, many British traders still operating there abandoned southwestern trade to contest the trade of Rupert's Land. For instance, Forsyth, Richardson and Company, a subsidiary of the London firm Phyn, Ellice and Company, which had operated in the southwest since 1783, turned from the trade of the United States to the northwest. This company merged with the Detroit firm of Leith, Jamieson and Company in October 1798 to form the New North West Company, which became better known as the XY Company (XYC).[3] Similarly the Montreal firm of Parker, Gerrard and Ogilvy, which had traded only on the south shore of Lake Superior before 1796, was in the Athabasca country by 1800. The worst fears of the NWC partners were realized in 1800 when this firm joined the XYC to form a formidable opposition to the NWC. Alexander Mackenzie, who originally apprenticed with Gregory, McLeod and Company, never felt at home in the NWC and apparently collaborated with the New North West Company even before his contract expired in late 1799. He joined the company officially in 1802, and it was reorganized as Sir Alexander Mackenzie and Company in 1803.[4]

Unlike the HBC, the XYC challenged the NWC not only in the Saskatchewan basin but also in the Athabasca. Between 1798 and 1804 spiteful personal animosities deepened the bitter and violent commercial rivalry between the two companies.[5] The competition had consequences for everyone. In their efforts to win the trade of the Indians the NWC and XYC brought about twice as much alcohol into the northwest in the 1802–3 season as the NWC had four years before.[6] Both companies operated at a loss. The cutthroat competition might not have ended until one of the companies collapsed, but the sudden death of Simon McTavish, preeminent figure in the NWC since 1779, smoothed the way for a merger on 5 November 1804.[7] The NWC partners were determined not to face another Montreal rival. The ferocious and violent northwestern trade had never been a place for the faint of heart, but after 1804 the NWC ruthlessly crushed all Canadian upstarts.[8] Now the struggle was between the NWC and the HBC, which appeared weak and passive compared to the flamboyant and aggressive NWC. The HBC had made only halfhearted forays into the Athabasca. It sent its first employee there in 1791 and established its first post there in 1802. Even that effort was abandoned in 1806.[9] The HBC still earned large profits during the years of NWC and XYC competition and small profits for several years thereafter.[10]

In the Saskatchewan River basin the HBC and NWC, spurred by a growing rivalry and by the depletion of fur resources near older posts, built many new posts that reached the foothills of the Rocky Mountains by the turn of the nineteenth century. The NWC took the initiative; but the HBC, which all but conceded the Athabasca to the NWC, competed effectively on the northwestern plains. In the spring of 1795 Angus Shaw of the NWC decided to build a post near the Beaver Hills, about five days' journey above Fort George / Buckingham House. Duncan McGillivray of the NWC explained that the post would be built at "the termination of an extensive plain contained between two Branches of this [Saskatchewan] River." He noted that

> this is described to be a rich and plentiful Country, abounding with all kinds of animals especially Beavers & Otters, which are said to be so numerous that the Women & children kill them with Sticks and hatchets.—The Country arround Fort George is now entirely ruined. The Natives have already killed all the Beavers, to such a distance that they lose much time in coming to the House, during the Hunting Season. The Lower Fort [Fort George / Buckingham House] will only therefore serve in future for the *Gens du Large,* whilst the Crees Assiniboines, and Circees [Sarcees], being the Principal Beaver Hunters will resort to the Forks [Augustus-Edmonton].—This division of the Indians will be doubly advantageous to the Co, both with respect to augmenting the usual returns & taking the [beaver-hunting] natives out of the reach of any opposition, (except the English) for the ensuing Winter at least.[11]

During the summer of 1795 James Hughes built Fort Augustus near the mouth of the Sturgeon River at present-day Fort Saskatchewan, Alberta, and William Tomison built the HBC's Edmonton House the following October. McGillivray's hopeful prediction that other Canadians would be unable to compete was wrong, however. Two small Canadian companies joined the NWC and HBC during the same autumn.[12]

After the XYC emerged in 1798, the HBC and NWC actually worked together to force the upstarts out of business.[13] HBC and NWC posts proliferated as the two companies tried to take advantage of the XYC's shortage of personnel.[14] The Rocky Mountain House/Acton House complex and the Chesterfield Houses were the most important of these new posts.[15] Traders established Rocky Mountain House and Acton House in the fall of 1799 at the mouth of the Clearwater River, about fifteen difficult days' travel up the North Saskatchewan River from Fort Augustus. While Edmonton sat on the edge of the plains, Rocky Mountain House lay in the foothills, which were "in general covered with wood and at intervals small prairions until about a mile or in some places more within, where large open swamps are found. The Wood principally is of Pine, of several kinds, Aspen, Willow and some Birch."[16]

The traders took a risk when they established posts at Chesterfield House. Built at the confluence of the Red Deer and South Saskatchewan Rivers in 1800, those posts were far from Cree and Assiniboine bands. Until 1800 the traders built all of their posts in or near the parkland region, but Chesterfield House was deep in the plains. The NWC partners probably devised the plan to build there, but the HBC traders actually arrived first.[17] Peter Fidler's crew constructed and named Chesterfield House in the autumn of 1800; Pierre Belleau of the NWC and John Wills of the XYC soon joined them. Fidler and Belleau built their posts within one stockade and, in keeping with their policy of noncooperation with the XYC, left Wills to build separate facilities.[18] In 1801 only the HBC and XYC returned to the site. Given the harrowing events of the 1801–2 season, it is surprising that the traders returned at all, but they abandoned the site only temporarily. In autumn of 1804 John McDonald and James Hughes led a crew of about twenty-five NWC men to reoccupy the posts.[19] Joseph Howse headed the twenty-man HBC brigade at the same site.[20] But the winter of 1804–5 seems to have been as tumultuous as the 1801–2 season, so the traders deserted the site for almost twenty years.

Why did the traders ever build at Chesterfield House? Ironically, their reasons probably had little to do with the resources of the plains and much to do with broader strategy. Successful exploration was crucial to the NWC. In contrast to the HBC, Canadian companies had always relied on costly explorations as they tried to outdistance their competition. The NWC had sponsored important explorations in the 1780s and 1790s, including Alexander Mackenzie's first known crossing of North America in 1793. But the NWC had failed to find a practical transportation route to the Pacific through the subarctic. If they could find such a route, they might yet nullify the advantages that the HBC held by owning exclusive rights to transportation in Hudson Bay. Having exhausted possibilities in the north, the NWC probably hoped the South Saskatchewan River system would enable them to open trade with the transmountain bands or even to develop a serviceable transportation route to the Pacific. In 1800 Chesterfield House also fit the NWC's aims of stretching the XYC's resources. Traders expected to gather many furs and provisions at Chesterfield House, and they did, but the NWC probably hoped that their post might become a link in a new east-west transportation corridor.[21]

Trading establishments were only semipermanent. Traders deserted poorly situated posts very quickly, but they also eventually abandoned or relocated well-situated posts when local fuel supplies and fur resources diminished. In 1800 the traders abandoned the Fort George/ Buckingham House complex and replaced it with Island House about twenty miles upstream.[22] They abandoned Island House after only one year because of its inconvenient location and built the Fort Vermilion/ Paint River House complex downstream in about 1802. The traders also left the original Fort Augustus and Fort Edmonton in 1802 and replaced them with new posts with the same names near the site of present-day downtown Edmonton.

As before, the spread of Euroamerican traders affected American Indian bands in important ways. Traders had their hopes for the new trading centers, and the Indians had theirs. Ultimately the posts became what the most powerful local bands wanted them to be. For example, in 1795 traders contrived to send the plains bands to Fort George and Buckingham House while the new Fort Augustus and Edmonton House served Cree, Assiniboine, and Sarcee beaver hunters exclusively. Piegan, Sarcee, Blood, Siksika, and Gros Ventre bands immediately began trading at the upper posts, however. The Blackfoot bands brought in a large part if not most of the trade at Fort Augustus and Edmonton House in the first few years, and traders soon stopped trying to send Blackfoot traders to the lower establishments.

The plains bands also interfered with traders' efforts to cater to the Kutenais. After several failed attempts Kutenai bands first visited trading posts during the late 1790s. At Fort George on 28 February 1795 Duncan McGillivray reported that

> the Coutonées a tribe from the Southwest are determined to force their way this year to the Fort or perish in the attempt: rumour reports that their Chief has got a parchment Roll written by the Spaniards to the traders of this quarter, the contents of which are unknown.—The Gens du Large and all the other nations in this neighbourhood wishing to retain an exclusive trade among themselves, have hitherto prevented the Intentions of this Band, of commencing a friendly intercourse with the Fort, in order to exclude them from any share of our commodities, which they are well aware would put their enemies in a condition to defend themselves, from the attacks of those who are already acquainted with the use of arms. . . . it is reported that they intend obtaining a safe passage hither by bribing their enemies with Bands of Horses. Whether this method will succeed we cannot judge, but it is shrewdly suspected that a party will be formed to intercept as usual their progress to this quarter.[23]

It seems McGillivray's hunch was correct, because records never mention a Kutenai visit to Fort George/Buckingham House. The first Kutenai visit to Euroamerican trading centers came in 1798 at Edmonton House. In March 1798 William Tomison greeted a large band of Bloods and Piegans who arrived with "two of the Cuttencha tribe which I sent for. Those have not brought any furs of any kind but by their account their country abounds with all kinds, but far off."[24] To encourage these Kutenais Tomison gave them gifts worth 40 MB, but the Kutenais must have been disappointed to see their Blood and Piegan hosts leave the post with thirty-six new guns. The Kutenais probably never visited the traders at Edmonton again. It was very dangerous for them to try to reach Edmonton without Blackfoot approval, and if they visited Blackfoot bands their hosts could easily exact whatever payment they wished. Considering the distances they had to travel, the Kutenais probably found the trips impractical. Fortunately for the Kutenais, the traders wanted to deal directly with them because their territory was believed to abound with beavers.[25] That is why the traders built Rocky Mountain House.

The companies established posts at Rocky Mountain House despite the apparent misgivings of some of the Blackfoot bands and the opposition of some western Cree bands.[26] The traders intended to open regular trade with the Kutenais, although they doubted from the beginning whether the post was far enough upstream to achieve the purpose. Because of the difficulties of traveling on the river above and below Rocky Mountain House, however, they judged it impractical to establish posts farther upstream. The local Piegans agreed to try to bring in some Kutenais, but they did not keep their promise.[27] The Piegans probably did not oppose the construction of the trade center because they understood that the posts were convenient for them but not for the Kutenais.

At first the traders tried to develop Rocky Mountain House into a Kutenai trading center, but uncooperative Piegans and Crees thwarted their plans. In October 1800 NWC trader David Thompson hired Nappé Kigow, a Piegan chief, and Eapis, a hunter, to help him find some Kutenais and bring them to Rocky Mountain House. With five NWC men and the two Indian guides, who probably went along more to monitor Thompson than to assist him, Thompson set out for the headwaters of the Red Deer River. Thompson's party intercepted twenty-six Kutenai men and seven Kutenai women in the rugged upper Red Deer River valley on 14 October. After the Kutenais presented gifts to Thompson and complained that the Piegans had stolen most of their horses, Thompson convinced them to visit Rocky Mountain House.[28] The visit was a failure. By the time the Kutenais arrived at Rocky Mountain House on 20 October the Piegans had taken most of their horses. Thompson applauded their resolve and bravery in the face of the far larger Piegan band, but it is unlikely that the Kutenais made the trip to prove their determination and courage.

In November Thompson and Duncan McGillivray visited Sakatow's Piegan band at Spitcheyee (near present-day High River, Alberta). Sakatow complained about the trade that had taken place between the NWC and the Kutenais at Rocky Mountain House,[29] but he must have understood that this post posed little threat to his people. Rocky Mountain House was impractical as a Kutenai trade center. The Piegan bands themselves traded primarily at this post and ensured that Kutenais rarely visited there, although over the years some Kutenais did come to trade. In the autumn of 1806 a Kutenai band traded eighty beavers at Acton House. Late in the same winter another five Kutenai men made an arduous sixteen-day trip over the Rockies carrying about 100 MB worth of furs on their backs. Heartened, James Bird asked J. P. Pruden, in charge at Acton House, to stay there for the summer in case he could arrange a rendezvous on the western side of the mountains with any other party of Kutenais that might visit.[30]

Bird's instructions show that the traders understood that any substantial trade with the Kutenais would have to take place at a location other than Rocky Mountain House. They had conceded that Rocky Mountain House and Acton House had become exactly what the Piegans intended them to be: a Piegan trading center. The traders may have been disappointed, but the fact that the western Piegan bands turned out to be the most prolific beaver hunters among the Blackfoot bands must have consoled them. There was nothing to mitigate the disappointment for the Kutenais. In 1806 they were as vulnerable on the northwestern plains as they had ever been.

The traders always expected the Siksikas to be their main clients at Chesterfield House. This post was a remarkable experiment, because the traders had to understand from the outset that they could not depend upon the Crees and Assiniboines for any assistance there. The Crees must have accepted Chesterfield House because they were still friendly with most of the Blackfoot bands. The Blackfoot bands must have accepted the idea because they could regulate the trade there. And the traders must have accepted the risk because they believed they could trust certain Siksika bands. Siksika bands led by Old Swan almost certainly decided where Chesterfield House would be built. Any knowledge Euroamericans might have had of the South Saskatchewan River beyond the old South Branch House had been lost, so no Euroamerican could have chosen the location of the posts.[31] Members of Old Swan's band also provided all the essential services that Cree and Assiniboine bands provided at the North Saskatchewan River posts. They were hunters, guides, interpreters, couriers, and horse herders.[32]

Even though the traders expected the cooperation of Old Swan's band, they never were at ease at Chesterfield House. On their initial ascent of the South Saskatchewan River they feared an attack. They were similarly apprehensive on their way downstream after their first trading season, even though it had passed peacefully.[33] They were anxious because the Gros Ventres were some of the main customers and Indians often vastly outnumbered traders at the post.[34] On 2 October 1801 Fidler estimated that 1,400 Indian men, women, and children were camped at the posts at Chesterfield House. This is far more people than would gather at North Saskatchewan River posts. Because the Gros Ventres arrived in large numbers, the traders were always on alert.

Most visitors to Chesterfield House were Siksikas and Gros Ventres, who overwhelmingly brought in fox and wolf furs but virtually no beaver skins. If the traders hoped to trade with Crow or Shoshoni bands or to acquire many beaver furs, they were disappointed. Although some daring Shoshoni men raided nearby and the Siksikas brought some Shoshoni slaves to Chesterfield House, the Blackfoot bands would have obstructed Euroamerican trade with their enemies.

The Gros Ventres, however, did bring some of their Arapaho kin to trade at Chesterfield House. On 31 October 1801 a band of about twenty "Tatood Indians" arrived at the post. Fidler learned that they were

> part of a nation that never saw Europeans before. They inhabit on the eastern borders of the mountain far to the south of this, they have been forty-four days in coming, they speak nearly the same language as the Fall Indians and are at peace with them, who have escorted them here. They brought a few beaver skins with them but very ill coloured and badly dressed, they are a pretty numerous tribe amounting to about 90 or 100 tents. Their manners are different from the Fall Indians, but are nearly of the same size and features.[35]

The Arapahos, who had arrived in the South Saskatchewan basin during the summer, stayed with the Gros Ventres until the spring. They probably never visited NWC or HBC posts again.

As Euroamericans moved onto the northwestern plains some Indians and non-Indians from within the region and from elsewhere seized opportunities to serve as intermediaries between trapping bands and the Euroamericans and as independent trappers. The Canadian companies always used *coureurs de drouine* (itinerant traders) who led trading parties from trading posts to Indian winter camps. These men, by marrying eligible women of such bands, formed kin relations with prominent adult Indian males. Since officers had little work for the many unskilled men at trading posts during the winter, they encouraged *engagés* (contracted employees) to spend much of the winter away from the posts, among their Indian kin in order to reduce costs and to establish and strengthen ties between indigenous bands and their particular trading concerns.

When their contracts expired, engagés, typically debt-ridden, faced an unpromising future in Canada. Some of those who were skilled hunters and trappers and were well connected with Indian bands chose not to return to Montreal but to remain in the interior, where they could use their subsistence, trapping, guiding, and interpreting skills and where they enjoyed some standing among both traders and indigenous bands. Kinship ties with Indian bands were crucial for these *hommes libres* (freemen), for they ensured that the freemen were part of the network of reciprocal obligations between related bands. Their position as intermediaries in the trade was also important.

Their unique economic and social role as intermediaries between indigenous bands and Euroamerican traders and as assiduous trappers encouraged many freemen families to intermarry with one another rather than with Indian or Euroamerican communities. That is how distinct mixed-blood communities emerged in the northwest.[36] Many of these proto-metis bands lived in the forests, but by the 1790s some bands with connections to local Cree bands wintered on the northwestern plains between the North Saskatchewan and Red Deer Rivers.[37] Freemen rarely had close ties to the Blackfoot or Gros Ventre bands and thus were limited to the northern reaches of the plains, where Cree and Assiniboine bands predominated.

After about 1800 many eastern Indians from communities near Montreal arrived in the northwest. Their canoeing, hunting, and trapping skills made them very attractive to trading companies. In the first few years of the nineteenth century the NWC brought hundreds of eastern Indian engagés to the North Saskatchewan River region.[38] Some northwestern indigenous groups were initially quite willing to accept the newcomers. For example, in November 1800 David Thompson and Duncan McGillivray got permission from the Piegans to take some Iroquois and Saulteaux trappers into the foothills of the Rocky Mountains.[39] Others, like the Gros Ventres, viewed them as a threat and were instantly hostile. Many eastern Indians quickly became freemen because they had the required subsistence and sociopolitical skills and were quickly integrated into the bands of other freemen.[40] Within a few decades Iroquois freemen were found throughout western North America.

Hostilities that flared up between Blackfoot bands and Cree and Assiniboine bands in the 1780s apparently subsided in the 1790s, but their former symbiotic relationship disappeared forever. Skillful diplomats and kinship ties prevented descent into war in the 1780s, but some of the leaders best able to broker peace agreements died in the early 1790s. Young Man, who had grown up among the Crees but became a prominent Piegan leader, had been a crucial link between Cree and Piegan communities for decades. Because of his background and demeanor Young Man "was universally beloved by all the Pecanow Tribe—& made himself respected amongst the adjacent friendly nations." In the spring of 1792, however, he was bitten by a beaver; "this bite not being properly attended, the warm weather coming on & being an old man—this mortified & carried him of[f] in June 1793."[41]

Young Man's death impaired the Piegan bands' ability to deal with Cree bands. Other Piegan leaders, like Sakatow and Kootenay Man, did not have the close family ties to the Cree bands that Young Man did. Sakatow, apparently born around 1740, was a handsome and well-built man with an outstanding war record. As a sign of his influence he carried two otter skins covered with mother of pearl. According to Thompson, many of the Piegans considered him too violent and quarrelsome, but he was a very able orator.[42]

More conciliatory, Kootenay Man (Kootanae Appe) was a very tall, thin man who appeared to have disdained and avoided Sakatow but was "both loved and respected, and his people often wished him to take a more active part in their affairs but he confined himself to War, and the care of the camp in which he was, which was generally of fifty to one hundred tents."[43] As the most respected warrior among the Piegans in the 1780s and 1790s Kootenay Man could typically lead bands farther southwest than any other Blackfoot bands. He was preoccupied, as were the Piegans generally, with neighbors to the south. So Kootenay Man had little influence over Blackfoot relations with the Crees. Especially after the death of Young Man, the Siksikas, who were neighbors of the Cree and Assiniboine bands, were most important in managing the relationships among these bands.

Leadership of the Siksikas also changed in the 1790s. Old Swan (or the Swan) was apparently the most influential Siksika leader during the 1780s, when the symbiotic relationships between the Blackfoot bands and the Cree and Assiniboine bands crumbled. McGillivray noted that Old Swan was "respected and esteemed by all the neighbouring tribes." By the 1790s, however, Old Swan, like Young Man, was old and frail, and in December 1794 he "unluckily Stumbled over a Dog, and broke some blood vessel which occasioned his death 2 days thereafter."[44] Thereafter the Siksika bands differed in their policies toward their neighbors.[45]

According to McGillivray, as Old Swan grew older and less vigorous Big Man gradually emerged as the foremost Siksika leader.[46] Big Man and Old Swan are a study in contrasts. Aptly named, Big Man was combative.[47] McGillivray commented that "this formidable cheif is universally feared by all the neighbouring nations, his immense size contributes Greatly to this distinction & some acts of personal courage which he has displayed on many occasions have established his reputation so firmly that he is supposed to be the most daring and intrepid Indian in this Department."[48] Big Man was among the Siksikas who attacked traders in 1793 and 1794. By then he had already killed the brother of Sitting Badger, a prominent Plains Cree leader and firm friend of NWC traders, who sought to avenge that death. Big Man led western Siksika bands that seemed to be closely associated with neighboring Blood bands. The traders certainly used the same disapproving words to describe the Blood bands and Big Man's bands.[49] Separated from each other by most of the Siksika bands, the Cree and Assiniboine bands and the Blood and Big Man's bands had little need to appease each other. Big Man was poorly positioned and little inclined to strengthen the friendships that survived among the Siksika and Cree bands by 1794. If anything, Big Man undermined the peace.

Big Man's Siksika rival was Old Swan's son, "Feathers," who symbolically assumed his father's name when his father died.[50] After 1794, although perhaps not before that date,[51] the younger Old Swan worked to preserve peace between the Blackfoot bands and the Crees, Assiniboines, and Euroamericans. Old Swan led the northeasternmost Siksika bands, who were most vulnerable to any Cree or Assiniboine attacks. Whether or not Old Swan's band had intermarried with Cree and Assiniboine bands, Old Swan was instrumental in keeping the peace between the Blackfoot bands and the Cree-Assiniboine bands until 1806.

Hostile incidents continually threatened to destroy the Blackfoot coalition with the Cree and Assiniboine bands between 1794 and 1806. Bands that normally camped far apart clashed most often, especially when Cree and Assiniboine men raided Blackfoot horses. The Crees and Assiniboines, particularly the Assiniboines of the northeastern plains, still held their reputation as the most prolific horse raiders in the region.[52] Between 1794 and 1806 traders on the North Saskatchewan River rarely mentioned hostilities between the Blackfoot bands and the Crees and Assiniboines, however.[53] Given that Indian warfare affected traders directly, they would probably have noted such warfare if it had been frequent. Evidence that Siksika trading parties often arrived at trading posts in small groups reinforces the impression that war was infrequent. Siksika bands evidently often wintered near trading posts and even north of the North Saskatchewan River. Meanwhile some Cree bands wintered at bison pounds south of the river.[54] Cree and Siksika bands could not mingle in the parkland region to this extent unless they were generally friendly toward each other.

Blackfoot and Gros Ventre paths clearly diverged in the 1790s. Despite occasional hostilities, most of the Blackfoot bands were usually at peace with the Crees and Assiniboines and had easy access to trading posts. Their only real enemies were their militarily weak neighbors to the south. In contrast the Gros Ventres and the Crees and Assiniboines hated one another bitterly.

While the Blackfoot and Gros Ventre bands grew more distant, the Sarcee and Blackfoot bands grew closer. The Sarcees had recently left the southern boreal forests, where beavers were abundant. In the 1780s traders praised the Sarcees, who hunted the forest-plains ecotone, as proficient beaver hunters. That changed when the Sarcees moved increasingly onto the plains by the 1790s. Sarcee bands pounded bison as far south as the Battle River by 1793.[55] In early 1798 William Tomison complained that a certain Sarcee band that arrived at Edmonton House "brought but little provisions and not above 200 wolves, and those very Indians used to bring above 1000 parchment beaver."[56] This Sarcee band was still linked to Cree bands, but as the Sarcee bands became equestrian plains bison hunters some of their bands also gravitated toward the Blackfoot bands.

We cannot explain relationships between the plains bands and the Euroamerican traders adequately except in the context of broader political, economic, and military trends on the northwestern plains. The instability of the northern coalition affected traders. Plains bands that had the most difficulty with the Crees and Assiniboines were also the most aggressive toward the traders. Traders understood this fact acutely after 1794; but although they tried to broker peace, they failed to influence relations among Indian bands significantly.

When plains Indians first met Euroamericans they were apparently awed. The Blackfoot bands coined their word Napikwan (Oldman Person) to refer to the newcomers, which suggests that they initially connected the Euroamericans with Napi, the creator, although it does not mean that they worshiped these men.[57] When the elder Old Swan died in 1794, McGillivray reflected that "his intentions towards the white people have been always honest and upright, and while he retained any authority his band never attempted anything to our prejudice."[58] Over time, however, plains Indians became less deferential. During the 1790s the Blackfoot bands still believed that the Euroamericans possessed exceptional powers, but they evidently doubted any connection between the Napikwan and Napi. The plains Indians knew how important Euroamericans were as arms brokers, although they also soon understood that the traders themselves were relatively powerless. Most plains bands and Euroamerican traders remained aloof from one another. Daniel Williams Harmon expressed the traders' perceptions well, commenting that "those Indians who reside in the large Plains are the most independent and appear to be the happiest and most contented of any People upon the face of the Earth. They subsist on the Flesh of the Baffaloe and of their Skins they make the greater part of their Clothing, which is both warm and convenient."[59] Relatively independent and self-reliant, plains bands did not defer to the traders.

For their part, Euroamerican traders were indifferent to plains Indian trade when compared to the trade of "thickwoods Indians." The European market for wolf skins was far less lucrative than the market for beaver and fancy furs and grew even less profitable over time. The HBC, which was willing to pay 2 MB for a prime wolf skin until 1789, dropped its standard to 1/2 MB in 1800. At the same time the Governor and Committee appealed to its traders to trade as few wolf furs as possible. The HBC changed its standard for good reasons; high import duties and low prices in Europe made it impossible to profit from wolf furs.[60] Plains bands still resented the changes.[61]

Relations between traders and Blackfoot and Gros Ventre bands grew less friendly in the 1780s and 1790s. Curiously, the Blackfoot and Gros

Ventre bands were far less likely to raid the traders' horse herds than were the Crees and Assiniboines (particularly the Assiniboines). Violent incidents occasionally marred the relationships between traders and Cree and Assiniboine individuals. Because the Cree and Assiniboine bands provided many valuable furs and essential services, however, traders overlooked difficulties with the Crees and Assiniboines more readily. Many traders spoke Cree and Assiniboine, and many married Cree and Assiniboine women. Blackfoot and Gros Ventre women often slept with traders when they visited the posts, but they rarely intermarried with them. Few traders understood the Blackfoot language, and none learned Gros Ventre.[62] Traders rarely gave goods on credit to Blackfoot and Gros Ventre men, something they did routinely to their Cree and Assiniboine clients.

After the troubles of 1793 and 1794 traders were cautious with the plains bands. When the HBC built Edmonton House in 1795, they equipped it with a musquetoon as a matter of course.[63] But if the traders grew more suspicious of the plains bands, the feelings were mutual. In 1803 Daniel Williams Harmon noted that the Gros Ventres, who considered the Crees and Assiniboines their enemies, "look upon us as there enemies also, for furnishing the Crees & Assiniboines with fire arms, while they have few or none, having as yet had but little intercourse with the White People."[64]

Plains bands and individuals understood how to make the most of their peripheral position in the trade. Because they lacked kin among the traders and were not indebted to them, plains bands were also free of reciprocal obligations that tied many other bands to specific traders. Bands that intermarried with traders often traded with their kin. Also, in order to avert disastrous competition, traders agreed at least in principle that they should not consent to trade with individuals and bands that were indebted to other traders. None of this applied to most plains bands, who were absolutely free to trade wherever they wished. Unconstrained, plains bands grew more aggressive at trading posts in the 1790s. Big Man's Siksika bands illustrate this. These bands complained and bargained relentlessly, intimidated traders, and stole goods outright to lower the costs of acquiring Euroamerican goods.[65] Trade occurred, but often in an atmosphere of conflict and defiance.

Some more conciliatory Blackfoot bands like the Siksika bands led by Old Swan did cooperate with the traders, perhaps because these bands resided near trading posts and Cree and Assiniboine bands. Old Swan's

Siksika bands dealt with traders and with Cree and Assiniboine bands with remarkable mutual trust. Whatever the reason for these different approaches, the defiance of some Blackfoot and Gros Ventre bands and the cooperation of others directly affected the development of trade on the northwestern plains. The services of Old Swan's bands made the establishment and operation of Chesterfield House possible, but the resistance of other bands led to its abandonment.[66]

Traders wanted peace among the Indians of the northwestern plains, so they tried to present themselves as peace brokers. Peter Fidler explained why the traders disliked warfare among their clients. After learning that young Siksika men had left on one of their frequent war expeditions, he predicted that "the above circumstance will be a great loss in trade at this house this year as the young men kill nearly all the furs the Indians have to trade. The old men does nothing but kill meat and spend their time in the tents in play and other acts of indolence, and cannot persuade themselves to be the least industrious."[67] Traders could not impose peace upon belligerents, but they did help broker agreements among bands that were amenable to peace.

Trading establishments seem to have been neutral ground by general consensus. Fratricidal violence was not rare at posts, but interethnic battles were. Indians sometimes sent peace overtures to other communities through trading establishments. Traders embraced their role as mediators by relaying these messages. When Indians told the traders that they intended to attack other bands, the traders implored them not to. The traders also often warned bands that they felt were in danger of attack. Euroamerican traders tried to remain neutral when their clients fought. They never fought on behalf of any Indian group. That certainly helped them avoid becoming too deeply embroiled in the escalating warfare. Significantly, at least some Indians accepted the traders as legitimate intermediaries, and Indian bands rarely attacked their enemies at trading posts.

Blackfoot power peaked between the 1794 and 1806, when peace between the Blackfoot and the Cree and Assiniboine bands held and the northern bands still prevented significant trade between Euroamericans and their militarily weak southern neighbors. The Blackfoot bands raided their southwestern neighbors with near impunity. In 1809 Alexander Henry the Younger explained that the Blackfoot "so eas[il]y prey on their enemies that many of the old men among the Blackfeet [Siksikas] have killed with their own hands during their younger days from 15 to 20 Men.... He is considered but a moderate Warrior that had killed but 10 Men."[68] Ambitious young men needed a respectable war record. Henry noted that a man's status depended on the number of scalps he had taken. According to Peter Fidler, those who had never killed an enemy were "looked upon by their Country men little better than Old Women."[69]

Between 1800 and 1805 the Blackfoot bands needed to travel no farther north than Chesterfield House or Rocky Mountain House to trade with Euroamericans. They used these posts as staging grounds for wintertime raids upon their enemies. Traders were still erecting the stockades of Chesterfield House in the fall of 1800 when fifty Siksikas assembled there to wage war on the Shoshonis. Forty more warriors passed through on 20 November as the HBC traders were laying the guardroom floor of their post. By the time Fidler reported that "we have entirely finished our house" on 1 December nearly all of the Siksika young men had gone to war against the Shoshonis.[70] The Chesterfield House journals for the 1800–1801 and 1801–2 winters show that the Blackfoot bands often raided toward the south. Given that summer was the principal season for warfare, this level of warfare in winter is remarkable.

Maps drawn by members of Old Swan's bands and by a Gros Ventre man depict the northwestern plains at the time (see figures 7.2 to 7.5). They show that the Blackfoot bands were familiar with the entire northwestern plains and beyond, particularly to the southwest and west. The Gros Ventres were more oriented toward the south: the Gros Ventre map sketched the western plains as far south as New Mexico. Old Swan's map of 1801 identifies rivers and features south of the Missouri River by their Gros Ventre names, suggesting that the Blackfoot bands had only recently become familiar with that region with the help of the Gros Ventres. The maps also reinforce the impression that Blackfoot dominance was expanding southward. The accuracy of the Siksika maps and the Blackfoot raids to the very southern limits of the northwestern plains testify to the military dominance of the northern coalition.

Old Swan's 1801 map also reveals that many of the southern bands had fled the plains at the turn of the nineteenth century. His map shows that the Blackfoot warriors harried the Shoshoni bands even in the upper Missouri basin beyond the open plains and in the valleys west of the

Rocky Mountains. Old Swan traced the path of a Siksika war party that left Chesterfield House on 22 November 1800 and returned on 10 January 1801. That expedition searched unsuccessfully beyond the three forks of the Missouri for Shoshoni victims. Old Swan's map indicates that the Crow and Arapaho bands wintered in the upper Musselshell basin and in the middle reaches of the Yellowstone River near the Bighorn Mountains. It shows "Mud House" people (either Hidatsas or Crows) near the confluence of the Judith and Missouri Rivers,[71] but it clearly illustrates that the vast region between the South Saskatchewan River and the Missouri River (and beyond) was not permanently occupied by any bands but was dominated militarily by the Blackfoot bands.[72]

Even as the southern bands endured repeated enemy raids, they faced another disaster: the return of smallpox. The previous epidemic struck every indigenous community on the northern plains, but the epidemic of 1801 spared the Blackfoot, Cree, and Assiniboine bands. So smallpox again weakened the military position of the southern coalition compared to the northern. Like the previous epidemic, the disease spread from south to north on the Great Plains. It may have originated in Mexico, although the Arapahos contracted it from other Indians who traded with Euroamericans on the lower Missouri River.[73] The Arapahos, in turn, passed the disease to their Gros Ventre kin while they visited them in the South Saskatchewan River basin in the summer of 1801. The epidemic reduced the Crows to about 2,400 people living in 300 tents.[74] The Crows also carried smallpox to the Flatheads.[75] From there the disease seems to have spread westward into the Columbia basin, but not northward beyond the Gros Ventres. The epidemic struck the vulnerable communities on the northwestern plains and spared the powerful.

Smallpox must have spread so readily among the southern bands because they often interacted. They were united by more than their hatred and fear of the Blackfoot bands. The southern bands were diverse, but differences helped tie them together. The Siouan-speaking Crows, for example, had kin among the Hidatsas. The Numic-speaking Shoshonis had important links with their kin who traded with the New Mexicans. Through fellow Salishans and other western Indians the Flatheads had access to goods from as far away as the Pacific Ocean. With links in many directions, the southern bands maintained their long-standing and significant trade among themselves. They were on generally good terms and probably intermarried often.[76] These groups clashed on occasion: Shoshoni bands raided Crow bands in 1793, and the Gros Ventres joined Crow raids on the Shoshonis. In 1802 some Crow bands raided the Flatheads, taking scalps and prisoners.[77] Still, the southern bands evidently maintained generally peaceful relations that tied them together symbiotically.[78]

Although the Flatheads had no direct access to Euroamerican traders, they had much to contribute to the trade network that linked the southern bands. The mild climate and abundant grasses of the Bitterroot Valley area made that region a horse nursery. The Flatheads traded their surplus horses and other products, including bows made from deer antlers and dentalium passed to them from the Pacific coast. In return the Crows supplied the Flatheads with goods they got at the Mandan/Hidatsa villages.[79]

The Shoshonis got Spanish goods via their kin who "have dealings with the white[s] of New Mexico from whom they get thick striped *Blankets*, Bridles & Battle axes in exchange for Buffaloe robes and Deer skins," but "what they get in this way is very trifling and cannot answer their purposes."[80] The Shoshonis traded some of these goods (like blue beads) as well as goods (like shells) acquired from their western neighbors and resources from their own territories (like obsidian) to their neighbors on the northwestern plains.[81]

By 1800 the Crows had the most valuable outside trading connections of the southern bands. Although the Mandans and Hidatsas were decimated by disease and war by 1800, their importance to the residents of the northwestern plains increased as Euroamerican goods became more readily available at their villages.[82] The Crows were important traders at the middle Missouri villages. Every year they sold many of the horses they got from the Flatheads at these villages "at duble the price they purchase them and carry on a continual trade in that manner."[83] The Hidatsas also sought the leather goods, robes, dried meat, and slaves that the Crows brought. In return the Crows received Euroamerican and Hidatsa products like guns and ammunition, kettles, axes, corn, pumpkins, and tobacco. The Crows also brought some Shoshoni and Flathead traders to the middle Missouri villages, just as they were escorted by the Shoshonis to trade fairs to the south.[84] The southern bands combined had extensive and valuable trading networks, an abundance of horses, and excellent Indian weaponry. Their crippling difficulty lay in their shortage of Euroamerican metal and weaponry.

Because they had the most dependable access to European weaponry, the Crows dominated the southern bands at the turn of the nineteenth century.[85] Unlike the other southern bands, the Crows hunted the northwestern plains all year. They were able to continue to live in the abundant Yellowstone Valley, because they were better armed than the other southern bands.[86] Although they had fewer horses than the Flatheads, their herds were more than adequate. By 1805 most Crow men owned at least ten horses and many had thirty or forty, which they tended carefully. They were also known as excellent horsemen.[87]

In the space of only a few years the Crows became much more capable of defending themselves. When Charles Le Raye visited the Crows in the upper Yellowstone and lower Bighorn Valleys in 1802, he reported that many of them had guns but no ammunition.[88] According to François-Antoine Larocque, who visited them in the same area only three years later, the Crows had long refused to trade guns and ammunition with their neighbors, "but this year as they have plenty they intend giving them some."[89] Their arsenal then consisted of bows and arrows, lances, war clubs, shields, and guns. Larocque considered the Crows poor marksmen with guns because they had been unable to practice with their scarce ammunition, but he regarded them as excellent marksmen with bows and arrows. Still able to hunt their favorite lands, the Crows rarely went to war except to defend themselves. Larocque believed that the Crows, having been decimated by the epidemic of 1801, camped together for protection when they feared an attack by their northern neighbors. According to Charles McKenzie, however, the Crows in 1805 consisted of two groups of bands, the Kegh-chy-sa (led by Nakesinia or Red Calf, who hosted Larocque in 1805) and the Hey-re-ro-ka (led by Red Fish). Combined, he believed they could raise a formidable six hundred warriors.[90]

Dominant on the northwestern plains only sixty years earlier, by 1800 the Shoshonis lived precariously along the mountainous margins of the northwestern plains.[91] By the beginning of the nineteenth century they were considerably weaker than the Crows. According to Le Raye, "this nation resides principally on the headwaters of the Big Horn river, and in the most inaccessible parts of the Rocky mountains, where they have frequently to hide in caverns from their enemies. Owing to their defenseless situation they become an easy conquest to any nation disposed to attack them, and they are frequently attacked for no other reason than the pleasure of killing them."[92] Le Raye described the "Snake" arsenal in 1802 as consisting of war clubs, shields, ten-inch bone daggers, and small bows.[93] Some Shoshoni men did own Euroamerican weapons. A small Shoshoni war party that attacked some Siksikas in 1802 had two HBC guns manufactured in 1790 and enough shot and powder to use the guns. They wounded one of the Siksika men before their nearly inevitable defeat.[94] The Shoshonis were still a poor match for the Blackfoot, Cree, or Assiniboine bands.

Poorly armed, the Shoshoni bands retreated to the salmon realm or the grassy valleys on the margins of the northwestern plains for much of the year. In August 1805 the Lewis and Clark expedition met a Shoshoni band in "extreem poverty" as they crossed from the Lemhi River into the Missouri basin. These Shoshonis, Lewis explained, subsisted on salmon but

> as this fish either perishes or returns about the 1st of September they are compelled at this season in surch of subsistence to resort to the Missouri, in the vallies of which, there is more game even within the mountains. here they move slowly down the river in order to collect and join other bands either of their own nation or the Flatheads, and having become sufficiently strong as they conceive venture on the Eastern side of the Rockey mountains into the plains, where the buffaloe abound. but they never leave the interior of the mountains while they can obtain a scanty subsistence, and always return as soon as they have acquired a good stock of dryed meat in the plains; thus alternately obtaining food at the risk of their lives and retiring to the mountains, while they consume it.[95]

Unlike the Crows, the desperate Shoshonis launched raids against the Blackfoot bands even into the Saskatchewan basin as late as the beginning of the nineteenth century. For instance, in the winter of 1801–2 the Siksikas discovered a party of Shoshoni horse raiders just a day's journey south of the Cypress Hills. During the same winter the Siksikas killed eight members of a party of ten Shoshoni horse raiders near the South Saskatchewan River only forty miles from Chesterfield House.[96] Why did the Shoshonis take these risks? Perhaps older Shoshoni men and women who remembered the years of Shoshoni dominance spurred and shamed their young men to action. Or perhaps their situation was so desperate that they felt had few other options. In any case the Blackfoot bands showed little fear of the Shoshonis at the turn of the nineteenth century.

The Flatheads, who lived in the mild and open environment of the Bitterroot and Flathead Valleys, had abundant horse herds, but their territory supported few if any bison. The bison herds that had lived west of the Rocky Mountains in the pedestrian era were extirpated by the nineteenth century.[97] Because many bands depended on the bison of only a small area along the southwestern margins of the plains, the bison were depleted there. The herds of the mountainous upper Missouri River basin diminished noticeably around the turn of the nineteenth century.[98] When they were west of the mountains, the Flatheads depended on deer and fish, primarily salmon. Perhaps the extermination of the bison herds west of the divide led Flathead bands to risk hunting on the plains. In the early nineteenth century they often crossed the mountains in the fall to hunt bison in the upper Missouri basin, to trade with other southern bands, and even to steal horses from or attack vulnerable Piegan encampments.[99] The Flatheads owned few Euroamerican goods before 1805; their weaponry consisted of bows, often made of antlers, and projectiles that were primarily bone tipped, although they probably fashioned some metal tips from worn metal kettles traded from the Crows.[100]

The Flatheads rarely raided the Piegans, but their northern neighbors, the Kutenais, never did. By the turn of the nineteenth century the Kutenais were a small and militarily weak group residing nearly permanently west of the Rocky Mountains. With only very tenuous access to Euroamerican goods, their trips to the plains were fraught with danger. By 1800, when two Canadian traders visited them, the Kutenai bands who formerly lived east of the Rocky Mountains wintered in the open country of the upper Kootenay River region. From there it was a demanding and dangerous seventeen-day journey to Rocky Mountain House. The Kutenais lived primarily on deer, which were common but difficult to kill. The Canadian visitors reported that the Kutenai band frequently went hungry; during the winter "they passed over the mountain to kill Buffalo there & remained 14 Days killing a sufficient stock—while they were doing this necessary business, some kept watch on the adjoining hills in case of a surprise by the muddy river Inds who are at this time declared & inveterate enemies." The Kutenais traveled between the Rocky Mountain trench and the northwestern plains less often than in the past. After many years of disuse their trails had become "very much encumbered with Wind fall wood &c." The mild winters of the Rocky Mountain trench supported many feral horses, but the Kutenais owned relatively few because horses were of little use to deer hunters. Neither did their lands support many beavers.[101]

The Hidatsas were important primarily because they supplied goods to the southern bands, but they also made forays onto the northwestern plains. By 1800 they appear to have been generally friendly with their Crow kin; these friendly relations and their access to Euroamerican weaponry evidently enabled them to continue traveling to the upper Missouri basin even when Cree and Assiniboine bands prevented them from traveling to the Assiniboine River basin.[102] The Hidatsas fought with the Blackfoot bands on the northwestern plains but also apparently with the Shoshonis. In 1805 the Hidatsa chief One Eye (Le Borgne) boasted that the Hidatsas took many scalps from the defenseless Shoshonis.[103] Hidatsa war parties may not have traveled to the northwestern plains often, but their visits had some very important consequences for the history of the region. Once, probably in 1800, a Hidatsa war party attacked a Shoshoni band at the three forks of the Missouri, killing at least ten people and taking several captives.[104] The captives were taken to the Hidatsa villages, where one of them, Bird Woman (Sacajewea), eventually married the Canadian-born Toussaint Charbonneau. While she was still a teen, Bird Woman, her husband, and her two-month-old baby accompanied Lewis and Clark on their famous expedition. As interpreter she helped establish peaceful relations between Lewis and Clark and the hard-pressed Shoshonis.[105]

The Hidatsas also delivered the death blow to HBC and NWC efforts at Chesterfield House. A party of Hidatsas came to Chesterfield House at Christmas 1804 to announce that they wanted to trade at that post. A Blackfoot band, however, detected and attacked them the next day.[106] It may be this incident that the Hidatsa leader Wolf Chief referred to when he told Charles McKenzie that his brother had been killed in the fall of 1804 on a war expedition against the Blackfoot bands. In March 1805 the Hidatsas sent a party of fifty men to avenge the defeat. Either because they could not find the Blackfoot bands or because they found a band too large to attack, the Hidatsa warriors turned back without engaging the enemy.[107] On their way home they discovered NWC traders from Chesterfield House, who on their trip down the South Saskatchewan River had stopped to rendezvous with some plains Crees near the Moose Woods, about halfway between Chesterfield House and the forks of the Saskatchewan. The

Hidatsas "attacked at day break, a volley was fired in Bouché's tent, where three men were in bed asleep, and all of them were killed."[108] The Hidatsas returned to their villages with several scalps, the possessions of the dead traders, half a keg or more of gunpowder, and at least 200 balls: "more," according to Larocque, "than ever I saw in the possession of Indians at one time."[109] The Hidatsas traded some of these goods with the Crows.

Unwittingly the Hidatsas had struck a serious blow to the Siksikas. The vulnerable NWC and HBC traders at Chesterfield House had already suffered the depredations of the Gros Ventres during the brief tenure of that post, and this Hidatsa attack appears to have convinced them to abandon Chesterfield House. The bands of the northern coalition saw their most convenient trading post abandoned in the very year when the Crow bands left the middle Missouri villages exceptionally well supplied with European weaponry.

After Gros Ventre and Siksika bands attacked traders in 1793 and 1794, the Gros Ventres faced growing challenges, but not primarily because of the policies of the traders themselves. Euroamericans never got any significant revenge, if they wanted it. Intense competition prevented the traders from presenting a common front against the Gros Ventres. Without such unity traders sacrificed short-term profits if they refused to deal with the belligerents. So Euroamerican leaders and the Gros Ventres quickly resumed trading. In May 1795 the HBC Governor and Committee asked that "no means be lost in trying to reconcile those Natives with our Servants."[110] Big Man's Siksika band, which had joined the attack on Pine Island and Manchester House, returned to trading posts in October and November 1794. Big Man and his band endured a tongue lashing and then relinquished some of the horses they had taken, but they paid a remarkably small price for their assaults.[111]

Despite initial reports that some of the Gros Ventre belligerents had abandoned the Saskatchewan River region permanently, they also restored their trading relations. Some of them evidently did go south after the attacks, probably to visit their Arapaho kin, but they returned within a few years. They were far more hesitant to approach the traders than Big Man's Siksikas were; they did not visit posts again until the autumn of 1795. On 25 November 1795 the very Gros Ventres who had attacked Manchester House and South Branch House arrived at the half-constructed

Fort Augustus and Edmonton House to sue for peace. The events of that visit were apparently such a bitter disappointment to William Tomison that he omitted the details in his post journal of that year. When the band arrived, the irascible Tomison, whom John McDonald sarcastically described as "my old Friend," apparently decided the Gros Ventres should be punished severely for their attacks.[112] To that end he met with the NWC officers, James Hughes and John McDonald, explaining that "he could not receive them [the Gros Ventres] as friends." McDonald, therefore, visited the large Gros Ventre encampment

> & with an Interpreter told them of Mr. Thompson's [Tomison's] resolve. They loaded me with kindness & Buffaloe fur Robes—they had by this time pitched their tents. They told me they would willingly make peace & not molest the Hudson's Bay establisht,—but would trade all they had with me—& was glad that I met them without any fear of any harm—and I placed confidence in them.
>
> They accordingly came on & we made a good Trade. Mr. Thompson biting his fingers at the result.[113]

The rivalry among traders had prevented them from punishing attacks even of the magnitude of the South Branch House incident. The Gros Ventres reconciled with the HBC during Tomison's year-long furlough in England in 1796–97. On a rainy 14 December 1796 four hundred Gros Ventres arrived at Edmonton. HBC and NWC officers reprimanded them before agreeing to restore normal relations.[114] In this sense the traders' behavior conformed to old and established diplomatic practices on the northwestern plains.

The Gros Ventres would not restore relations with the Crees and Assiniboines so easily. The devastating attacks of the Crees and Assiniboines in the late 1780s and early 1790s had incited the attacks on traders in the first place. The plunder the Gros Ventres acquired in their attacks did not make them a formidable military power, and the Cree and Assiniboine bands' attacks on them appear to have continued without pause and probably intensified. In November 1794 McGillivray learned that Cree warriors were gathering to go "on an expedition against the Slave Indians."[115] While at Nipawin in January 1795 James Bird noted that the seventy tents of the Saskatchewan, Red Deer (the Red Deer River that flows into Lake Winnipegosis), Swan, and Red River Cree bands were pounding bison in preparation for a war expedition against the Gros Ventres.[116] In the spring of 1795 a Cree man was killed by the Gros Ventres near Buffalo Lake.[117] Repeated bloodshed did not augur well for Gros Ventre–Cree relations.

David Thompson summed up the history of the Gros Ventres in these years by suggesting that "their chief was of a bad character, and brought them into so many quarrels with their allies, they had to leave their country and wander to the right bank of the Missisourie, to near the Mandane villages."[118] "L'Homme de Callumet" (Pipe Man), one of the preeminent leaders of the Gros Ventres shortly before 1794, died in the South Branch House attack that year.[119] His brother A kas kin succeeded him as "great chief," but he and most of the other Gros Ventre leaders at that time were as bellicose as L'Homme de Callumet had been. In 1801 Fidler wrote of A kas kin that "his character relating to his regard for Europeans are none of the best but on the contrary he is generally stirring his countrymen up against them."[120] The only Gros Ventre leader who might have been inclined to make peace with the Crees and Assiniboines was Kate thak ki, but he does not appear to have sufficient influence within his own community to broker an agreement.

Realistically the Gros Ventres had few alternatives but defensive warfare. There was very little basis for a peace agreement between the Cree and Assiniboine bands and the Gros Ventres. On their own the Gros Ventres were too small and vulnerable to sue for any peace on equal terms. The traders, who fancied themselves peace brokers, must have been reluctant to speak on behalf of the Gros Ventres. Neither was there much reason for the Blackfoot leaders to throw their weight behind the Gros Ventres. Big Man's band may have sided with the Gros Ventres in 1793 and 1794, but the aftermath of the attacks appears to have undermined Big Man's influence even within his own bands.[121] Furthermore, Big Man was a warrior, not a diplomat. Old Swan's band, which apparently still enjoyed friendly relations with neighboring Cree and Assiniboine bands, was in the best position to intercede but, given the fragility of the peace that existed between the Blackfoot bands and the Plains Crees and Assiniboines, it is not surprising that the Gros Ventres would be on their own for the time being.

Without strong friends, the Gros Ventres withdrew toward the southwest after 1794. They would have sought the successful and long-lived

Chesterfield House even more than Old Swan's band did. Chesterfield House, situated near the Gros Ventre wintering grounds, made it unnecessary for the Gros Ventres to travel to North Saskatchewan River posts. They may very well have welcomed the establishment of Chesterfield House. According to Peter Fidler, the Gros Ventres consisted of about 180 tents with about 600 warriors at the time. Given their military weakness and their diplomatic isolation, it is not surprising that the Gros Ventres were exceptionally eager to maximize their returns. Fidler reported that they produced proportionately more furs (especially fox furs) than the Siksikas and prepared their skins with greater care.[122] The Gros Ventres appear to have behaved very peacefully during their first winter at Chesterfield House.

Any hopes that the Chesterfield House posts might have instilled among the Gros Ventres in 1800 proved to be illusory. The Gros Ventres suffered tremendous reversals during 1801. In February a party of Siksikas led by Old Swan's son killed two Gros Ventre men only a day's journey from Chesterfield House. They took 170 fox furs. In October Old Swan tried to appease the Gros Ventres by giving them two horses, but by that time far more serious disasters had befallen them. On 12 and 13 May 1801, after several weeks of springlike weather, a sudden snowstorm lashed parts of the northwestern plains while the Gros Ventres were in the upper South Saskatchewan Valley. Not far away at Moose Woods the weather was not cold, but snow fell "without intermission for 2 days & nights & the Snow on the level was knee deep, much more than we saw the whole Winter, and the drifts in some places was more than 6 feet deep."[123] That spring blizzard killed over a hundred of the Gros Ventre horses.

During the summer northeastern Cree and Assiniboine warriors killed at least seventy-four Gros Ventres in two devastating attacks in a region the Gros Ventres had long hunted. Forty-six died in the first attack in the Cypress Hills; twenty-eight more died in the second attack in late August on the upper Oldman River near present-day Lethbridge. In all the Gros Ventres lost fourteen men, sixty women and children, and well over a hundred horses. Tragically, smallpox hit them the same summer, carrying off at least a hundred of their young people.[124] It is no wonder that the traders found the Gros Ventres in a desperate mood in the fall of 1801. The traders greeted them with more bad news. The HBC lowered the standard on wolf skins to 1/2 MB in 1800; but word did not get to Peter Fidler until well into the trading season, and he apparently did not apply it during that season. Beginning in the autumn of 1801, however, traders at Chesterfield House accepted wolf furs at only 1/2 MB.[125]

More disasters hit in the ensuing autumn and winter. In October men of Big Man's band killed four Arapahos who were visiting with their Gros Ventre kin near Chesterfield House. Tensions between the Gros Ventres and Big Man's Siksika bands ran high all winter. On 27 January 1802 a sudden and ferocious winter storm struck, killing two Gros Ventre men who had become disoriented near the camp. Fidler described the day as "the *worst day* I ever saw in this Country." He reported: "Strong gales, with the very great drift & snow—could not see 100 yards."[126] The storm and the deep snow that it left behind claimed a lot of horses. Many of the Siksika horses died "on account of the great depth of snow, not being able to scrape the snow away and maintain themselves." Some of the HBC horses met the same fate.[127] After losing hundreds of horses during the previous year and eighty more in the storm, the Gros Ventres found their horse herds badly depleted.

Peter Fidler understood the implications of these catastrophes for the traders at Chesterfield House. Within days of arriving at Chesterfield House in 1801 he noted that "the Fall Indians, on account of the war and disease this summer cutting off such number of them, appears desperate, and is nearly ready to fall on anyone they can." Gros Ventre behavior foreshadowed a tense winter. They were combative during the entire 1801–2 trading season, but after the January blizzard Fidler wrote that "all these losses and misfortunes coming upon them so soon after each other has a great deal soured their dispositions, which before this late affair was nearly upon a balance whether to do good or bad."[128] The bloodshed began in February. Early on the morning of 21 February 1802 the men of the XYC post at Chesterfield House repulsed about seventy Gros Ventre attackers. The traders were not yet aware, however, that the Gros Ventres had already attacked and killed several Iroquois nearby.

Four Canadians and fourteen Iroquois working with the NWC traveled up the South Saskatchewan River during the winter of 1801–2 to trap in the Cypress Hills. On 19 February 1802 four of the Iroquois and two Canadians went ahead to greet the local bands. Old Swan's Siksika band welcomed them as friends, but on 19 February 1802 the Gros

Ventres killed two of the Iroquois near Chesterfield House. On the morning of 22 February the Gros Ventres killed the other two Iroquois after they left Old Swan's tent only four miles from Chesterfield House. They let the two Canadians go. The Gros Ventre scalped, dismembered, and mutilated the dead bodies.[129] Evidently they saw these trappers as interlopers who wanted to exploit the same source of furs that they relied upon.

Curiously, the traders at Chesterfield House traded with these Gros Ventres after they promised to let the other two Canadians and ten Iroquois reach the posts. The situation became even more tense when Big Man's band, which had avoided Chesterfield House since their run-in with the Gros Ventres and Arapahos in the fall, arrived on 1 March just as the entire Gros Ventre band, together with Arapaho guests, was camped at Chesterfield House. Fidler noted that the Arapahos were "all assembled about the House with guns to kill some of the Big Mans gang in retaliation for the man & woman and 2 children last fall."[130]

Big Man's band hastily traded their few furs and left, but in the early morning of 3 March 1802 two hundred Gros Ventre men attacked and killed the two Canadians and ten Iroquois sixteen miles from Chesterfield House. Again they dismembered the bodies, but this time they paraded past the posts displaying the scalps on the end of poles and under their belts and threatening to treat the traders the same way. The thirty-seven people in the HBC and XYC posts then burned the XYC post and joined together within one stockade. They did not try to avenge the killings because they knew they would be easy targets once they embarked downstream in the spring.[131] Instead Fidler tried to intimidate the Gros Ventres by drawing upon his friendships with Siksika and Piegan bands (some of whom arrived on 19 March) and by spreading rumors that the Crees and Assiniboines were planning to attack them in the summer. These measures apparently had the desired effect. Little Bear (Ki oo cuss), a Siksika leader Fidler trusted, who had tried to mediate between the traders and the Gros Ventres,[132] soon reported that "every Fall Indian with their families are pitched away for the Mis sis su rie River." Little Bear also warned the traders, however, that the Gros Ventres hoped "to meet the Crow Mountain and Tattood Indians, and make a very formidable party, and that they will then come to fall upon us and proceed down the country to find the Crees and Stone Indians and kill what they can of them in revenge of the last summer."[133] After the experiences of the 1801–2 season and the warnings of the intentions of the Gros Ventres, none of the trading companies appear to have returned to Chesterfield House until the autumn of 1804.

The possibility that the Gros Ventres and Arapahos might join with the Crows to attack the Crees and Assiniboines illustrates the Gros Ventres' tenuous yet flexible position between the northern and southern coalition. It also shows the convoluted paths that warfare and diplomacy took on the northwestern plains. The Gros Ventres had generally friendly relations with Old Swan's band at Chesterfield House between 1800 and 1802, even as they bristled at Big Man's band. So the Gros Ventres had good relations with the very Siksikas who were at peace with their Cree and Assiniboine foes but were at odds with the Siksikas who were hostile to the Crees and Assiniboines. During those same winters, however, Gros Ventre men fought the Shoshonis, both by themselves and in company with Crow war parties. The Gros Ventres brought at least two captured Shoshoni girls to Chesterfield House after combined Gros Ventre–Crow raids upon the Shoshonis.[134] In the spring of 1805 a Gros Ventre party attacked Cameahwait's Shoshoni band near the three forks of the Missouri, killing twenty, taking several more prisoners, and capturing many horses and all the Shoshoni lodges.[135]

As hard-pressed as the Gros Ventres were in these years, they and their Arapaho kin clearly had a unique ability to join with bands from the Yellowstone basin to the North Saskatchewan basin. In August and October 1802 Charles Le Raye met "Paunched Indians" on the foothills of the upper Yellowstone River. According to Le Raye, these Indians, many of whom had guns with no ammunition, lived on the upper Yellowstone and the Bighorn Rivers. He clearly described them as separate from the Gros Ventres (his term for the Hidatsas), so these were certainly Gros Ventre–Arapahos.[136] In September 1805 Larocque witnessed a wary but peaceful meeting between a group of Gros Ventres and a Crow band at the Yellowstone River near the site of present-day Billings, Montana. Some Shoshonis were also present. The Gros Ventres he saw represented a band of apparently 275 or 300 lodges encamped in the Big Horn Basin. According to Larocque, they "brought words of peace from their nation and say they came to trade horses. They were well received by the [Crow] Indians and presents of different articles were made them, they told me they had traded last winter with Mr. [John Mc]Donald [at Chesterfield House] whom they made known to me a[s] crooked arm."[137]

Any rapprochement between the Gros Ventres and the Crows would have done nothing to soothe Gros Ventre relations with the Cree and Assiniboine bands. Peace grew more elusive. On 18 August 1803 Daniel Williams Harmon at Fort Alexandria in the Assiniboine River basin noted that forty lodges of Crees and Assiniboines with others had set off in the spring on a war party but had separated at the Battle River (half a month's journey) "to go and make Peace with their Enemies, the Rapid & Black Feet Indians, for both parties begin to be weary of such a bloody War as has for such a length of time been kept up between them, and are therefore much inclined to patch up a Peace on almost any terms whatever." But less than three years later, while at South Branch House, Harmon noted that "the greater part of our Indians have gone to wage War upon the Rapid Indians, their inveterate enemies—with whom they often patch up a Peace, but is never of a long duration."[138] Seven days later he reported that the Gros Ventres had killed several Assiniboines within fifteen miles of that fort.

By 1806 the northern coalition was permanently fractured. The Gros Ventres occupied a nebulous middle zone, enjoying peaceful relations with some Blackfoot and Crow bands and with their Arapaho kin. Had peace between the Blackfoot bands and the Crees and Assiniboines survived much longer, the Gros Ventre might well have been driven from the Saskatchewan River basin and affiliated with the southern coalition. The remnant northern coalition, however, was fragile. Symbiotic trading relations between Blackfoot bands and Crees and Assiniboines had evaporated in the 1780s. Peace survived thanks to established kin connections and friendships among the bands and the efforts of Indian diplomats and Euroamerican peace brokers, but irritants constantly weakened the relationship. Repeated Cree and Assiniboine raids on Blackfoot horse herds caused most of the friction. At the beginning of 1806, however, the Blackfoot-Cree-Assiniboine-Sarcee peace was intact.

Meanwhile the southern coalition faced the repeated assaults of their powerful northern enemies. The southern bands were united by trading networks and by their common dread of the northern bands. By 1806 many of them hunted the bison of the northern plains only seasonally and at considerable risk. The Crow bands were the most secure of the southern bands. They still hunted the abundant Yellowstone River basin, but their hunting parties feared their well-armed foes. At the beginning of 1806 the southern bands could have harbored little hope that their situation would soon improve. Before the end of the year things changed dramatically.

Epilogue

Point out to them [the Blackfoot bands] as forcibly as possible the necessity there is of their being on a friendly footing with the Southward [Cree] Indians for to have a safe & easy Intercourse with us and above all that if they value our Friendship & assistance it will always be necessary for them whatever Quarrels may arise among the Indians, to consider us a party unconcerned, Friends to all, sorry for their dissentions and at all time willing to do everything in our power to compose them.

—JAMES BIRD TO J. P. PRUDEN
28 January 1807

A party of Americans were seen last Summer where the Missoury enters the rocky Mountain & this resisted by the Muddy or Missoury River Indians that four of them set off with an intention to come here but that they killed one & the rest returned.

—JAMES BIRD TO JOHN MCNAB
23 December 1806

The Blackfoot bands were at the center of the dramatic changes that made 1806 a turning point in the history of the northwestern plains. In the Saskatchewan basin the long-standing but fragile northern coalition dissolved. While acrimonious incidents had punctuated the general amity among the Blackfoot, Cree, and Assiniboine bands before 1806, and various bands managed to maintain peace for several more years, the normal state of affairs from 1806 to 1870 would be a Cree and Assiniboine coalition against the Blackfoot, Gros Ventre, and Sarcee bands.[1]

The realignment that began in 1806 affected everyone on the northwestern plains. Perhaps in the very month when the northern coalition disintegrated the Piegans encountered a small party of Euroamericans in the upper Missouri River basin. The Piegans learned the unsettling news that the Euroamerican newcomers had already made peace with their enemies, the Shoshonis and Flatheads. In the next year the flow of European goods to the southern coalition from both the Saskatchewan River and the Missouri River increased substantially, exactly as Lewis and Clark had predicted it would. The shifting alignments and dramatic change in the balance of military power on the northwestern plains promised to affect every inhabitant and even the environment.

The northern coalition disintegrated during the summer of 1806 when a large combined party of about four hundred Siksikas, with some Bloods, and about four hundred Crees and Assiniboines "were on their way to wage war on the Rapid Indians, their common enemy"; but along the way a battle broke out between the two groups, apparently arising from a dispute over the ownership of a horse.[2] The altercation left twenty-eight of the Siksikas and three of the Crees dead and forced the Crees "to a precipitate retreat with a loss of part of their Horses & baggage and dispersing in all quarters to conceal themselves in the woods leaving their Enemies masters of the Plains from South Branch to Acton House." The Siksikas, meanwhile, threatened "indiscriminate vengeance."[3] Their opportunity came in September when a group of two or three hundred Siksikas met a small band of about twenty-five Crees who were returning from a visit with the Piegans, unaware of the events of the summer. The Siksikas attacked the Crees about a hundred miles from Edmonton, killing or enslaving all but two men. The warfare of the summer of 1806 foreshadowed the continual warfare that marked the relationship between a new central coalition (composed of Blackfoot, Gros Ventre, and Sarcee bands) and a northern coalition (composed of Plains Cree and Plains Assiniboine bands).[4]

The years surrounding 1806 were a turning point in the Missouri and Columbia River basins as well. Crow and Shoshoni bands had hosted Canadian traders on the northwestern plains before 1806, but the Canadians could never make effective use of the Missouri River. At the beginning of 1805 the southern bands could not have expected the situation to change. By the end of 1806, without exception, the bands of the southern coalition anticipated direct trading relations with Euroamerican traders or trappers. In 1805 and 1806 Shoshoni and Flathead bands met members of the Lewis and Clark expedition, envoys from the United States, who promised that traders would soon arrive to supply them with the weapons they desired. The Crows did not meet these envoys but established some contact with United States fur traders in 1807. In 1807 NWC traders established direct trading relations with the Kutenais.[5] Within a few years the Euroamerican presence spread along the borders of the northwestern plains, and the southern bands' access to Euroamerican goods became far more reliable. The history of the northwestern plains had entered a new era.

Conclusion

The history of the northwestern plains is ancient, dynamic, complex, and fascinating. For centuries it was the common and contested ground of various human societies. During the bison era the plentiful bison herds of the region repeatedly drew communities from adjacent but less abundant environments. Developing sophisticated subsistence strategies and complex seasonal rounds, these diverse communities frequently met each other in peace and in warfare. Despite enormous cultural differences among them, some developed long-standing symbiotic relationships. Culturally dissimilar communities occasionally mingled, merged, fused, and even formed new "mixed" identities. Elsewhere members of a single culture could take different paths: they sometimes split, diverged, and even developed new identities. Members of each band on the northwestern plains sought their own security in constantly changing circumstances.

The arrivals of the bow and arrow, the horse and gun, Old World diseases, and Euroamerican traders were among the important milestones in the history of the northwestern plains between A.D. 200 and 1806, but the advent of the horse and gun between 1730 and 1770 was probably the most dramatic turning point in this entire period. The chronically uneven distribution of horses thereafter encouraged the development and entrenchment of two interethnic coalitions of bands. The southern coalition of Crow, Shoshoni, Flathead, and Kutenai bands was wealthy in horses but poor in guns, while the northern coalition of Blackfoot, Gros Ventre, Sarcee, and Cree bands had fewer horses but many guns. Between

1740 and 1750, as the distribution of guns and horses changed, the northern coalition rapidly moved from the defensive to the offensive.

When Euroamerican traders arrived, they became influential but not powerful participants in events in the region. Like previous newcomers, they found a place on the northwestern plains by forging partnerships with prior inhabitants. Euroamericans were no more united or monolithic than the indigenous bands. They, like other residents of the plains, competed among themselves for the resources of the region even as they established symbiotic relationships with the northern coalition of bands. The Euroamericans aspired to a position of neutrality from which they could broker peace agreements among indigenous groups. Their own self-interests determined this approach.

For Euroamericans, however, neutrality was an impossible dream. In practice fur traders were members of the northern coalition. They traded exclusively with these bands and developed particularly close relations with some of them. When forces unleashed by the presence of Euroamericans fractured the northern coalition, traders inevitably became embroiled in the conflicts that ensued. Euroamerican traders' response to the attacks of their own partners reveals how completely they had become integrated into the network and practices of trade, warfare, and diplomacy. Fur traders' efforts to restore relations with their trading partners conformed to the established conventions on the northwestern plains; they could not rely on their imported Euroamerican customs or standards of law, justice, or diplomacy. Given their limited power, Euroamericans enjoyed scant ability to direct the behavior of their partners. The fur traders' arrival and continued residence on the northwestern plains undermined the very alignment of bands from which the traders profited, and all the diplomatic efforts of the Euroamericans failed to prevent the coalition's gradual dissolution. The traders also could not stop their partners from waging unrelenting warfare upon their rivals.

The history of the northwestern plains is almost certainly not unique. If we assume that band-level or tribal societies were "primitive" or "simple," we will inevitably overlook the inherent complexity of all societies. If we reduce the relationships between representatives of Western societies and indigenous societies to their cultural dimensions, we cannot hope to reconstruct the past adequately. Much of the scholarship of the past century addresses the degree of cultural change and continuity in the history of indigenous societies. Many of the debates surround the cultural consequences of Euroamerican-newcomer contacts. These debates are not irrelevant, but they do not encourage the development of a literature that attempts to understand the past in its context. If the Indians of the northwestern plains experienced cultural crises upon the arrival of Euroamericans, the evidence for them is scant. The arrival of Euroamerican goods and Euroamericans themselves is vital to an understanding of aspects of this history, whether or not it brought dramatic cultural change. By moving beyond the limits of culturalist scholarship, researchers can ensure that the historiography of indigenous societies can remain the vital and innovative field it has been since the 1970s.

Notes

ABBREVIATIONS

GAI	Glenbow Alberta Archives
HBCA	Hudson's Bay Company Archives
NAC	National Archives of Canada

INTRODUCTION. BEYOND CULTURALISM: COMMUNITIES IN CONTACT

1. Lewis, *Effects*, quotations from pp. 3 and 61. Also see p. 5.
2. Sahlins, "Return of the Event," 44–45.
3. Tyrrell, ed. *Samuel Hearne*, 253.
4. See Helm, ed., *Essays*.
5. Sharrock, "Crees," 96.
6. See White, *Middle Ground*, xiv, for a discussion.
7. The following description of the band societies of the northwestern plains is the product of impressions gained from the documentary evidence but is also informed by the following important sources: Owen, "Patrilocal Band"; Helm, "Bilaterality"; Turnbull, "Importance of Flux"; Damas, *Contributions*; Savishinsky, "Mobility"; Ray and Freeman, *"Give Us,"* 14–18; June Helm, "Introduction," in Helm, ed., *Subarctic*; Dempsey, "Blackfoot Indians"; Albers, "Symbiosis"; Moore, "Putting Anthropology "; and Sahlins, *Tribesmen*, 20–21.
8. Moulton, ed., *Journals*, 5: 119–20.
9. Lamb, *Sixteen Years*, 215.
10. Moore, "Putting Anthropology," 935.
11. Owen, "Patrilocal Band," 675.
12. Sharrock, "Crees."
13. For example, Saukamappee (Young Man), born a Cree, became a prominent Piegan chief. Old Star, a prominent Plains Assiniboine, was born a Kutenai;

and Hugh Munro, born in Montreal, and Jimmy Jock Bird, of mixed English-Cree heritage, became prominent in Piegan bands as adults. Munro died on the Blackfeet Reservation in the 1890s. The ability of mixed-background individuals to serve as brokers between communities seems to have been persistent. During the transition from bison hunting to reserve life many prominent leaders were interethnic. Of the prominent American Indians of the treaty era, Big Bear was of mixed Ojibwa and Cree heritage; Minahikosis (Little Pine), born around 1830, was Blackfoot-Cree; Paskwa was a Cree-born leader of the Plains Saulteaux; Piapot was a Cree leader who lived with the Dakotas for much of his youth; and Poundmaker, a prominent Plains Cree leader, was the son of a Stoney and the adopted son of Siksika leader, Crowfoot. For other examples, see Albers, "Changing Patterns," 112–13. Richard White's *Middle Ground* (p. 392) notes that these permeable ethnic boundaries were also a feature of Indians in the *pays d'en haut.*

14. See Bruce Trigger's monumental *Natives and Newcomers,* especially his discussion on pp. 169–70.

CHAPTER 1. "A GOOD COUNTRY"

1. Bamforth, *Ecology,* 53–84. Most scholars now agree that the southern plains were not a particularly good bison habitat; Dan Flores, "Bison Ecology," 469–71.

2. Shannon and Smith, "Observations."

3. The Cypress Hills have a carrying capacity two to five times that of the surrounding plains; Vrooman, Chattaway, and Stewart, *Cattle Ranching,* 12; A. W. Bailey, personal communication, 28 October 1996. The Hand Hills produce about four times the forage produced by surrounding plains; A. W. Bailey, personal communication, 28 October 1996.

4. Vrooman, Chattaway, and Stewart, *Cattle Ranching,* 13.

5. Clarke, Tisdale, and Skoglund, *Effects of Climate.*

6. Black et al., *Effect,* 2.

7. Wilson, "Early Historic Fauna," 228; Umfreville, *Present State,* 153; HBCA B.60/e/ Fort Edmonton District Report, 1815.

8. Morgan, "Bison Movement Patterns," 149, 151.

9. Payne, *Vegetative Rangeland,* 6.

10. Coupland and Brayshaw, "Fescue Grassland," 390; Rowe and Coupland, "Vegetation," 238; and Watts, "Natural Vegetation," 36.

11. GAI M736, p. 23.

12. GAI M736, p. 28.

13. Dawson, *Report,* 10c–11c.

14. Spry, *Palliser Expedition,* 420, 19.

15. Quoted in McConnell, "Report," 10c.

16. Quoted in GAI M736, p. 32.

17. Dawson, *Report,* 17c.

18. Spry, *Palliser Expedition,* 406; also see pp. 18–19.

19. Black et al., *Effect,* 2; Payne, *Vegetative Rangeland,* 12.

20. Moulton, *Journals,* 4: 380.

21. Dawson, *Report,* 9c; Gordon, *Of Men and Herds,* 18; and Morgan, "Bison Movement Patterns," 154.

22. Morgan, "Bison Movement Patterns," 149. Morgan also argued that beaver dams stabilized water levels in many river valleys, enhancing the growing conditions for grasses and making water available to animals.

23. Horton, "Some Effects"; see especially pp. 278, 288–89.

24. Coupland and Brayshaw, "Fescue Grassland," 399; Wright and Bailey, *Fire Ecology,* 107; Smoliak, Willms, and Holt, *Management,* 17–19; Rowe and Coupland, "Vegetation," 238; Willms, Smoliak, and Dorman, "Effects." Important grasses that grow on the foothills and highlands but are not found in the northern parkland include the robust foothills rough fescue, Idaho (or bluebunch) fescue (*Festuca idahoensis*), and Parry oat grass (*Danthonia parryi*).

25. Clarke et al., *Ecological and Grazing Capacity,* 24–28. Also see Gordon, *Of Men and Herds,* 20; and Morgan, "Bison Movement Patterns," 148.

26. GAI M739, p. 31.

27. Smoliak, Willms, and Holt, *Management,* 13–14; Black et al., *Effect,* 2; and Clarke et al., *Ecological and Grazing Capacity,* 24; Wilson, "Early Historic Fauna," 228.

28. Horton, "Some Effects," 279, 95.

29. GAI M736, pp. 23–24. Also see GAI M736, p. 32.

30. See Longley, "Frequency of Chinooks."

31. Glover, ed., *David Thompson's Narrative,* 47.

32. Gough, ed., *Journal,* 20 January 1811; also p. 583.

33. Longley, *Climate,* figs. 4 and 5 (p. 12).

34. Clarke, Tisdale, and Skoglund, *Effects,* 24.

35. See Longley, *Climate,* figs. 4, 5, and 32 (pp. 12 and 45).

36. Johnston and Smoliak, "Reclaiming Brushland"; Fitzgerald and Bailey, "Control." Also see Smoliak, Willms, and Holt, *Management,* 17.

37. HBCA E.3/2, 18 December 1792; Raby, "Prairie Fires"; Rowe, "Lightning Fires"; and Nelson and England, "Some Comments," 295.

38. Wright and Bailey, *Fire Ecology,* 121; Campbell et al., "Bison Extirpation," 361. For wallowing, see Arthur, *Introduction,* 14–15.

39. Arthur, *Introduction,* 16; Fitzgerald and Bailey, "Control"; Fitzgerald, Hudson, and Bailey, "Grazing Preferences."

40. Gough, *Journal,* 32; also see p. 47 and HBCA B.239/a/69, 9 October 1772. For a discussion, see Bird, *Ecology,* 28.

41. Morgan, "Bison Movement Patterns," 150; Nelson, *Last Refuge,* 140.

42. Campbell et al., "Bison Extirpation," 360–62. Arthur suggested the importance of the bison in this regard in *Introduction,* 13–16. Also see Nelson *Last*

Refuge, 133. The role of the bison in maintaining short-grass prairie is discussed in Larson, "Role of Bison."

43. Arthur, *Introduction*, 30; Daubenmire, "Ecology," 242.

44. Courtney, "Pronghorn."

45. Pyne, *Fire in America*, 81. Also see Lewis, "Maskuta," 18; and Turner and Butzer, "Columbian Encounter," 18.

46. See Turner and Butzer, "Columbian Encounter"; Williams and Hunn, *Resource Managers*; E. Russell, "Indian-Set Fires"; Pyne, *Fire in America*, 45–51, 71–99.

47. Lewis, "Maskuta," 18; Pyne, *Fire in America*, 43.

48. Arthur, *Introduction*, 22–27; Lewis, "Fire Technology," 54–55; Nelson and England, "Some Comments," 297–98; and Loscheider, "Use of Fire." Also see HBCA E 3/2, January 1793.

49. Quoted in Arthur, *Introduction*, 25; also see pp. 26–27.

50. Ray, *Indians*, 133; Loscheider, "Use of Fire."

51. Vickers, "Seasonal Round Problems," 62.

52. Le Raye, "Journal," 177.

CHAPTER 2. THE ANNUAL CYCLE OF BISON AND HUNTERS

1. Gough, *Journal*, 535–36.

2. Roe, *North American Buffalo*, 674.

3. Roe, *Indian*, 197–99; quotation from p. 197.

4. Oliver, *Ecology*, 1–90.

5. Arthur, *Introduction*; Moodie and Ray, "Buffalo Migrations"; Morgan, "Bison Movement Patterns."

6. Hanson, "Bison Ecology."

7. Malainey and Sherriff, "Adjusting."

8. Hind quoted in Arthur, *Introduction*, 24.

9. Grinnell, *Blackfoot Lodge Tales*, 234.

10. Flores, "Bison Ecology," 476.

11. Grinnell, *Blackfoot Lodge Tales*, 240–41, 207.

12. Ewers, *Blackfeet: Raiders*, 86; Kehoe, "How the Ancient Peigans Lived."

13. Vickers, "Seasonal Round Problems," 60–62.

14. Ray, *Indians*, 75–78; Gough, *Journal*, 46.

15. Moulton, *Journals*, 8: 106; HBCA B.239/a/69, 9 October 1772; Abel, ed., *Tabeau's Narrative*, 71; Wissler, *Material Culture*, 41; Ewers, *Horse*, 152; and Ewers, *Blackfeet: Raiders*, 76.

16. Spry, *Palliser Expedition*, 258. Also see Gough, *Journal*, 284; and Moulton, *Journals*, 11 July 1806 (8: 104).

17. Gough, *Journal*, 15 July 1810; Morgan, "Bison Movement Patterns," 152.

18. Burpee, "York Factory," especially p. 331.

19. Ewers, *Blackfeet: Raiders*, 86.

20. Gough, *Journal*, 372; Verbicky-Todd, *Communal Buffalo Hunting*, 5; Frison, "Role of Buffalo."

21. Burpee, "York Factory," 340.

22. Gough, *Journal*, 284. Also see Burpee, "York Factory," 332.

23. Vickers, "Seasonal Round Problems," 64.

24. Vickers, "Seasonal Round Problems," 66.

25. Morgan, "Bison Movement Patterns," 153.

26. Horton, "Some Effects," 89.

27. Grinnell, *Blackfoot Lodge Tales*, 234.

28. Spry, *Palliser Expedition*, 266.

29. Tyrrell, *David Thompson's Narrative*, 338; Umfreville, *Present State*, 166.

30. HBCA B.239/a/69, 23 October 1772. William Pink's description is quoted in Russell, *Eighteenth-Century Western Cree*, 102. For another early description of a bison pound, see Bain, ed., *Travels*, 299–301.

31. Vickers, "Seasonal Round Problems," 64.

32. Wissler, *Material Culture*, 41. Also see Arthur, *Introduction*, 100, and Ewers, *Horse*, 152.

33. Hudson and Frank, "Foraging Ecology."

34. Christopherson and Hudson, "Effects"; Christopherson, Hudson, and Richmond, "Feed Intake"; Arthur, *Introduction*, 41–42.

35. Richmond, Hudson, and Christopherson, "Comparison"; Peden et al., "Trophic Ecology."

36. Moodie and Ray, "Buffalo Migrations," 49.

37. Arthur, *Introduction*.

38. Gough, *Journal*, 107 (14 January 1801).

39. Grinnell, *Blackfoot Lodge Tales*, 234.

40. Vickers, "Seasonal Round Problems," 65.

41. Verbicky-Todd, *Communal Buffalo Hunting*, 34–36; Arthur, Wilson, and Forbis, *Relationship of Bison*, 82.

42. Lamb, *Sixteen Years*, 42.

43. Gough, *Journal*, 374; Glover, *David Thompson's Narrative*, 51.

44. Spry, *Palliser Expedition*, 21.

45. Gough, *Journal*, 109, 115.

46. HBCA B.121/a/3, 24 May 1789. Also see B.121/a/4, 6 June 1789; and Morton, *Journal*, xlvii–xlviii.

47. Gough, *Journal*, 18 April 1810.

48. Grinnell, *Blackfoot Lodge Tales*, 234.

49. Speth and Spielmann, "Energy Source."

50. Arthur, Wilson, and Forbis, *Relationship of Bison*, 30.

51. Gough, *Journal*, 21 December 1809.

52. B.121/a/5, 19 February 1790. Also see B.121/a/4, 24 May 1790.

53. See B.121/a/5, 11 January 1790. Bulls, however, were harvested in winter to extract the fat from the marrow bones; B.121/a/5, 25 January and 5 March 1790.

54. Gough, *Journal*, 210 (18 July 1806).

55. HBCA E 3/2, 10 February 1793.

56. After the unusually cold and snowy winter of 1767–68 plains Indians told traders that there were few bison to be found and the ones that could be found were in poor condition.

57. Ray, *Indians*, ch. 2, deals with patterns of interaction in the parkland region north of the plains.

58. Henry found bison and bighorn sheep "very numerous" at the Kootenay Plains along the North Saskatchewan River in 1811; Gough, *Journal*, 506.

59. Hopwood, *David Thompson*, 224; Reeves, "Bison Killing," 78.

60. Butler, "Bison Hunting," 106–12. In 1792 the Piegans told Peter Fidler that there was "not a single Buffalo" west of the Rockies in the vicinity of the Oldman River basin: HBCA E.3/2, "Journal of a Journey," 31 December 1792.

61. Duke and Wilson, "Cultures," 63.

CHAPTER 3. TRADE, WARFARE, AND DIPLOMACY FROM A.D. 200 TO THE EVE OF THE EQUESTRIAN ERA

1. The "late pedestrian era" here refers to the period from the arrival of bow and arrow technology in roughly a.d. 200 to the arrival of the horse in roughly 1700, which archaeologists commonly refer to as the "late prehistoric era." The most valuable surveys of this era in northwestern plains history include Vickers, *Alberta Plains*; Walde, Meyer, and Unfreed, "Late Period"; various chapters (especially Vickers's) in Schlesier, *Plains Indians*; and Brink, *Dog Days*.

2. The best example of a romantic portrayal of indigenous warfare is Turney-High, *Primitive Warfare*, for many years the only general survey of warfare in nonstate societies. For the plains, see Grinnell, "Coup and Scalping"; and Smith, "War Complex." The best example of a realist interpretation of indigenous warfare is Keeley's *War before Civilization*. An early realist interpretation of plains Indian warfare can be found in Secoy, *Changing Military Patterns*. Also see Bamforth, "Indigenous People." A fine case study is found in Willey's *Prehistoric Warfare*, 152.

3. Bamforth, "Indigenous People"; Willey, *Prehistoric Warfare*, especially p. 152.

4. For a similar argument, see Trigger, *Natives and Newcomers*, 110.

5. Bamforth, "Indigenous People."

6. Teit, "Salishan Tribes," 306. This view is echoed in Blackfoot oral traditions described in Grinnell, *Blackfoot Lodge Tales*, 242.

7. Tyrrell, *David Thompson's Narrative*, 328–40.

8. Conner and Conner, *Rock Art*, 14, 16; Habgood, "Petroglyphs."

9. Keyser, "Plains Indian."

10. Conner and Conner, *Rock Art*, 14, 17.

11. Bamforth, "Indigenous People," 112.

12. Reeves, "Communal Bison Hunters," 184; Brink and Dawe, *Final Report,* 298.

13. Duke and Wilson, "Cultures," 60.

14. Brink and Dawe, *Final Report,* 296.

15. Reeves, "Communal Bison Hunters," 170.

16. Brink and Dawe, *Final Report,* especially p. 296.

17. Reeves, "Communal Bison Hunters," 170; Brink and Dawe, *Final Report,* 297–98, 302.

18. See Frison, *Prehistoric Hunters,* 223–24; Kehoe, "Paleo-Indian Drives," 82; and Reeves, "Communal Bison Hunters."

19. Kehoe, "Paleo-Indian Drives," 79. Also see Kehoe, *Gull Lake Site,* 195–96.

20. Reeves, "Communal Bison Hunters."

21. Archaeological cultures are defined differently than anthropological cultures. Archaeologists use the word "culture" to describe a set of related archaeological materials. Although scholars assume that there is usually some connection between archaeological cultures and ethnicity, they acknowledge that the relationship is not straightforward. Over the course of several centuries an unknown number of distinct communities may have left artifacts now categorized as Avonlea.

22. It appears that few now accept the earlier view of Reeves that the oldest Avonlea artifacts are from the western foothills; Vickers, "Cultures," 17. Note, however, that Duke and Wilson argue for the presence of Avonlea in the Rocky Mountain trench as early as a.d. 200; Duke and Wilson, "Cultures," 65. Greiser also noted the discovery of other very early Avonlea components in the foothills in Greiser, "Late Prehistoric," 41.

23. Reeves, *Culture Change,* 162–63.

24. Vickers, "Cultures," 16.

25. Walde, Meyer, and Unfreed, "Late Period." Other societies represented by the Pelican Lake phase, formerly widespread on the northern plains, also continued to survive in uplands of the upper Missouri and Yellowstone River basins until a.d. 1000; Greiser, "Late Prehistoric," 41.

26. Walde, Meyer, and Unfreed, "Late Period," 20.

27. Frison, *Prehistoric Hunters,* 223.

28. Brumley, *Ramillies.*

29. Meyer and Hamilton, "Neighbors," 108; Walde, Meyer, and Unfreed, "Late Period," 23.

30. David and Fisher, "Avonlea Predation," 101–18.

31. Vickers, "Cultures," 16.

32. This is the argument of Greiser, "Late Prehistoric," 42.

33. Vickers, "Cultures," 22.

34. Byrne, *Archaeology,* 469–70, 559; Walde, Meyer, and Unfreed, "Late Period," 22–23, 54.

35. Vickers, *Alberta Plains,* 106.

36. Hurt, *Indian Agriculture,* 62.

37. Hall, "Cahokia Identity," 33.

38. Two collections focus on the relationship between the Cahokians and their northern neighbors, but neither discusses relationships beyond the region of present-day South Dakota and Iowa: Stoltman, *New Perspectives*; and Emerson and Lewis, *Cahokia.*

39. Wood, *Papers,* 7–9; Tiffany, "Overview," 89–90.

40. Hall, "Cahokia Identity," 24–26.

41. Schlesier, "Commentary," 336; Gregg, *Overview,* 32.

42. Stewart, "Mandan and Hidatsa," 292; Russell, "Puzzle," in Epp, *Three Hundred,* 84.

43. Tiffany, "Overview," 107.

44. Wood and Downer, "Notes."

45. Nicholson, "Interactive Dynamics," 103–7; Nicholson, "Orientation."

46. Buchner, "Geochronology"; Nicholson, "Ceramic Affiliations."

47. Tiffany, "Overview," 89.

48. Gregg, "Archaeological Complexes," 88; Bamforth, "Indigenous People," 104–5.

49. Bamforth, "Indigenous People," 102–10; Willey, *Prehistoric Warfare.*

50. Walde, Meyer, and Unfreed, "Late Period," 33.

51. Meyer, "People," 64.

52. Meyer, "People," 64; Meyer and Epp, "North-South Interaction"; Meyer and Hamilton, "Neighbors," 127; Walde, Meyer, and Unfreed, "Late Period," 44.

53. Meyer, "People," 66.

54. Mulloy, *Hagen Site.*

55. Johnson, "Problem," 23.

56. Horse bones and fragments of brass or copper were found at the site; Forbis, *Cluny,* 16.

57. Byrne, *Archaeology,* 560.

58. Forbis, *Cluny,* 6, 67.

59. Forbis, *Cluny,* 72, 69.

60. Fidler quoted in Johnson, ed., *Saskatchewan River,* 266, 266n; emphasis in the original.

61. Le Raye, "Journal," 172.

62. Wood and Thiessen, *Early Fur Trade,* 197.

63. Vickers, *Alberta Plains,* 107.

CHAPTER 4. MIGRANTS FROM EVERY DIRECTION: COMMUNITIES OF THE NORTHWESTERN PLAINS TO 1750

1. Indeed the Blackfoot vocabulary is so dissimilar to that of other Algonkian languages that many prominent early linguists did not perceive the connection. The similarity becomes evident in the grammar. See Hale, "Report," 701.

2. Kroeber, *Arapaho,* 5; Kroeber, "Ethnology," 145.

3. The connection between Duck Bay and Blackduck is discussed in Meyer and Hamilton, "Neighbors," 120.

4. Kroeber, "Ethnology," 146.

5. Nicholson, "Ceramic Affiliations," 39; Malainey, "Gros Ventre," 31–32.

6. Mooney, *Ghost-Dance Religion,* 954.

7. Tyrrell, *David Thompson's Narrative,* 235–36.

8. See Milloy, *Plains Cree,* 5.

9. See Mandelbaum, *Plains Cree.*

10. Russell, "Puzzle," 84; Russell, *Eighteenth-Century Western Cree,* 210.

11. Wood and Thiessen, *Early Fur Trade,* 71, 74.

12. Abel, *Tabeau's Narrative,* 74.

13. Russell, "Puzzle," 78–79.

14. Burpee, "York Factory," 331.

15. Ray, *Indians,* 53; Russell, *Eighteenth-Century Western Cree,* 181–84.

16. Walde, Meyer, Unfreed, "Late Period," 49–50.

17. Epp, ed., "Henry Kelsey's Journals," 219. For Russell's interpretation of the evidence, see Russell, "Puzzle," 81.

18. Wood and Downer, "Notes."

19. Ewers, *Crow Indian,* 147.

20. Wood and Downer, "Notes"; Denig, *Five Indian Tribes,* 137; Bradley is quoted in Wood and Downer, "Notes," 89.

21. Teit, "Salishan Tribes," 304.

22. B.239/a/69, 1 December 1772.

23. For a defense and discussion of this theory, see Vickers "Cultures," 28–30.

24. B.239/a/2, 1 September 1716.

25. Ray, *Indians,* 55–57.

26. McKenzie quoted in Wood and Thiessen, *Early Fur Trade,* 249.

27. Suttles and Elmendorf, "Linguistic Evidence," 45, 41.

28. This evidence is discussed in later chapters.

29. Turney-High, *Ethnography,* 10; Teit, "Salishan Tribes," 304, 358. Also see Schaeffer, "Plains Kutenai," 2.

30. Chamberlain, "Report," 575.

31. Reeves, *Culture Change*; Vickers, "Cultures."

32. See Binnema, "Common and Contested Ground," 409–12.

33. See Ives, *Theory,* ch. 2, especially p. 14; Foster, "Language," 74.

34. Ives, *Theory,* 41–45. Also see Moodie, Catchpole, and Abel, "Northern Athapaskan."

35. Ives, *Theory,* 48–51.

36. Wilson, "Report on the Sarcee," 243; Jenness, *Sarcee Indians.* Jenness's report is based on fieldwork completed in 1921.

37. Jenness, *Sarcee Indians,* 3.

38. Shimkin, "Comanche-Shoshone Words," 198; Shimkin, "Shoshone-Comanche Origins"; Wright, "Shoshonean Migration"; Madsen, "Dating," 82–86; Bettinger and Baumhoff, "Numic Spread"; Sutton, "Warfare and Expansion"; and Bettinger, "How, When, and Why," 44–55.

39. See Greiser, "Late Prehistoric." For a dissenting interpretation, see Vickers, "Cultures."

40. The peaceful trade relationships also may have encouraged technological borrowing. The Crows appear to have adopted Shoshoni tri-notched points; see Greiser, "Late Prehistoric," 50.

41. See Madsen and Rhode, "Where Are We?"

42. Shimkin, "Shoshone-Comanche Origins," 20–21.

43. Butler, "Bison Hunting," 106–12.

44. Binnema, "Common and Contested Ground," 413–17. For a different interpretation, see Vickers, "Cultures," 29–30.

CHAPTER 5. THE HORSE AND GUN REVOLUTION, 1700–1770

1. Kroeber, "Ethnology," argues that the horse revolutionized plains cultures. Wissler, *Material Culture,* disagrees. Lewis's *Effects,* which directs attention toward the trade in furs, should be understood in the context of this debate. In the first in-depth study of the effects of the horse on a plains indigenous society Ewers tended to side with Kroeber; see Ewers, *Horse,* especially p. 338.

2. Secoy, *Changing Military Patterns,* 94. More recent work echoes Secoy's conclusion. See Flores, "Bison Ecology," 467; McCauley, "Conference Overview," 2; and Albers, "Symbiosis," 101–2.

3. Flores, "Bison Ecology," 467.

4. The most important studies that deal with equestrian transition and the arrival of the gun on the northwestern plains include Ewers, *Horse*; Lewis, *Effects*; Secoy, *Changing Military Patterns*; Milloy, *Plains Cree*; and Ray, *Indians.* Other worthwhile studies include the relevant parts of Fowler, *Shared Symbols*; Russell, *Eighteenth-Century Western Cree*; and Ewers, *Blackfeet: Raiders.*

5. The most detailed discussion of the founding of New Mexico is found in Simmons, *Last Conquistador.* For information on settlers, see p. 96. The classic study of the arrival of horses is Haines's "Where?" and his "Northward Spread."

6. Weber, *Spanish Frontier,* 77, 90, 196, 434n, 128.

7. See Weber, *Taos Trappers*; Thomas, *Plains Indians*; Hanson, "Mexican Traders."

8. See Haines, *Horses,* 51; Secoy, *Changing Military Patterns,* 20–30.

9. Rinn, "Acquisition," 22.

10. Secoy, *Changing Military Patterns,* 7.

11. Shimkin, "Introduction," 517.

12. Flores, "Bison Ecology," 468. Also see Secoy, *Changing Military Patterns,* 28; and Shimkin, "Comanche-Shoshone Words," 199.

13. Shimkin, "Introduction," 517; Sutton, "Warfare," 70.

14. Flores, "Bison Ecology," 468.

15. Shimkin, "Introduction," 517; Hanson, "Spain," 7; Arthearn, "Time of Transition," 19; Kavanagh, *Comanche,* 64–65.

16. Secoy, *Changing Military Patterns,* 29; Shimkin, "Comanche-Shoshone Words," 199. For a discussion of Comanche trade at Taos and Spanish-Comanche hostility, see Kavanagh, *Comanche,* 63–132.

17. Tyrrell, *David Thompson's Narrative,* 329.

18. Tyrrell, *David Thompson's Narrative,* 328.

19. Tyrrell, *David Thompson's Narrative,* 328–30.

20. Tyrrell, *David Thompson's Narrative,* 330.

21. Burpee, *Journals,* 412.

22. Hoxie, *Parading,* 43; Swagerty, "Indian Trade," 353.

23. Greiser, "Late Prehistoric," 50.

24. Mulloy, *Hagen Site*; Forbis, *Cluny*; Hoxie, *Parading,* 39; Vickers, "Cultures," 25.

25. Gough, *Journal,* 522.

26. Rinn, "Acquisition," 65–66, and Ewers, *Indian Life,* 12–13, argue that the Blackfoot bands would not have acquired their horses from the Shoshonis. It is less likely that they got them from the Shoshonis than from the Flatheads, but it was not impossible for such an exchange to have taken place between particular Shoshoni and Piegan bands that had agreed to a peace.

27. Fragments of metal, useful for arrowheads, were probably distributed more widely, although Young Man's band, resident on the lower Saskatchewan River, still used some stone arrowheads in the 1730s. See Tyrrell, *David Thompson's Narrative,* 328 (quoted above).

28. See Ray, *Indians,* 54, for a map showing the approximate distribution of European goods by 1720. It suggests that the hinterland of York Factory may have reached the northeastern part of the northwestern plains by 1720.

29. Douglas and Wallace, eds., *Twenty Years,* 40, 38; also see p. 42 and B.239/a/3, 11 June 1717.

30. B.239/a/2, 15 June 1716.

31. B.239/a/3, 11 June 1717, 28 May 1717; also see 25 June 1717; B.239/a/2, 18 June 1716; B.239/a/3, 11 March 1717 (quotation).

32. B.239/a/3, 10 June 1717.

33. See Innis, *Fur Trade,* 49–51. For the glut, see Eccles, *Canadian Frontier,* 125–26.

34. Eccles, *Canadian Frontier,* 145.

35. B.239/a/11, 12 June 1729; Innis, *Fur Trade,* 91–29; B.239/a/23, 4 June 1742.

36. The "Shusuanna" are mentioned in B.239/a/13, 24 June 1730; and the "Kis.ska.che.wan" in B.239/a/17, 31 May 1735.

37. James Isham to Ferdinand Jacobs, 17 July 1758, HBCA. Without explanation, Oscar Lewis estimated that the Blackfoot bands got their first guns in about 1728; *Effects,* 16.

38. B.239/a/17, 4 June 1735.

39. Innis, *Fur Trade*, 92.

40. Burpee, *Journals*, 250.

41. Meyer and Thistle, "Saskatchewan River," 418.

42. B.239/a/40, 22 July 1754; and Burpee, "York Factory," 352.

43. Ray and Freeman, *"Give Us,"* 34.

44. See Arthur J. Ray, "Wapinesiw," in *Dictionary of Canadian Biography*, 4: 761.

45. Burpee, *Journals*, 34; B.239/a/58, 3 May 1768 and 13 May 1768. "Earchethinue" and "Archithinue" are attempts by traders to write the Cree word for "Stranger," which apparently had some disparaging connotations and was sometimes translated at "Slave." It was a generic term by which Cree informants often referred to other groups. The traders eventually often used the term "Slave Indians" to refer to various Plains bands (usually the Blackfoot and Gros Ventre bands collectively). The names of Great Slave Lake and Lesser Slave Lake also have their origins in this Cree sobriquet.

46. B.239/a/69, 16 December 1772; B.239/a/36, 26 June 1754; Burpee, "York Factory," 346; B.239/a/63, 1 April 1770; B.239/a/56, 30 April 1767.

47. Milloy, *Plains Cree*, 6.

48. Tyrrell, *David Thompson's Narrative*, 330.

49. Tyrrell, *David Thompson's Narrative*, 330–32.

50. Tyrrell, *David Thompson's Narrative*, 335–36.

51. B.239/a/63, 1 April 1770.

52. In the version of the journals found in B.239/a/40 Henday specifically mentions meeting "bloody Indians" on 16 May 1755.

53. The label "Pegogoma" (muddy water) suggests that these Crees lived near the "Muddy Water River" (South Saskatchewan River) and the "Muddy Water Indians" (Piegans); Russell, *Eighteenth-Century Western Cree*, 142–43. Curiously, the HBC traders would later assume that the "Muddy Waters" were the Missouri River.

54. B.239/a/36, 22 June 1754, 26 June 1754, and 14 October 1754.

55. Burpee, "York Factory," 337–38.

56. Burpee, "York Factory," 338.

57. Burpee, "York Factory," 352.

58. Meyer and Thistle, "Saskatchewan River," 418; A. 11/115, fol. 37d.

59. Ray and Freeman, *"Give Us,"* 34.

60. James Isham to Ferdinand Jacobs, 17 July 1758, quoted in Ray, *Indians*, 55; James Isham and Council to Governor and Committee, 16 September 1758, A.11/115 fol. 16.

61. Rich, ed., *James Isham's Observations*, 113; Williams, ed., *Andrew Graham's Observations*, 257. The Made Beaver (MB) was a unit of currency equal in value to one prime beaver skin.

62. Ferdinand Jacobs to Moses Norton, 25 July 1763, B.239/b/24.

63. Andrew Graham and Council to Governor and Committee, 18 August 1766 and 18 August 1768, A.11/115.

64. B.239/a/69, 15 December 1772.

CHAPTER 6. THE RIGHT HAND OF DEATH, 1766–82

1. Crosby, "Virgin Soil," 299.

2. The scholarly literature on the history of the northwestern plains between 1766 and 1782 is small. Readers should turn to relevant portions of Ewers, *Blackfeet: Raiders*; Fowler, *Shared Symbols*; Lewis, *Effects*; Milloy, *Plains Cree*; Russell, *Eighteenth-Century Western Cree*; and Secoy, *Changing Military Patterns.* Although not focused on the northwestern plains, Ray's *Indians* is a very important study.

3. Ray and Freeman, *"Give Us,"* 34.

4. A. 11/115, Andrew Graham to Governor and Council, 26 August 1772.

5. Meyer and Thistle, "Saskatchewan River," 419. This is apparently the post that Cocking visited in 1772.

6. A. 11/115, Andrew Graham and Council to Governor and Committee, 18 August 1768.

7. This is shown in Meyer and Thisle, "Saskatchewan River."

8. A. 11/115, Ferdinand Jacobs to the Governor and Committee, 20 August 1768, (fol. 116d). Also see A. 11/115, Andrew Graham to the Governor and Committee, 26 August 1772 (fol. 142d); 30 August 1773 (fol. 161d).

9. Tyrrell, *Samuel Hearne,* 5–6.

10. A. 1/44, fols. 79–79d.

11. Rich, *Cumberland House* (1951), Cumberland House Journals, 13 May 1780; see pp. lxii–lxix. Rich explains that the canoe shortage was the most urgent concern at this time but that once this problem was solved in 1780 the British war efforts led to a chronic manpower shortage over the next decades.

12. Rich, *Cumberland House* (1951), 159, 115.

13. Bain, *Travels,* 235, 320.

14. W. McGillivray, NAC MG 19 B 4, section 10.

15. See Rich *Cumberland House* (1951), xliii, lv, and Hudson House Journals, 20 November 1779; Cumberland House Journals, 9 June 1777, 5 September 1777.

16. Rich, *Cumberland House* (1951), 323. Posts at or above the middle settlement were also known as "Forts des Prairies."

17. This incident must have been important to the traders' collective memory, for accounts of it are many. See Tyrrell, *David Thompson's Narrative,* 320, Morton, *Journal,* xxxiii–xxxiv; Gough, *Journal,* 361; Rich, *Cumberland House* (1952), Hudson House Journals, 4 March 1781. For a discussion, see Rich, *Cumberland House,* (1951), xlvii–xlix.

18. Rich, *Cumberland House* (1951), lii.

19. B.239/a/69, 1 December 1772.

20. Lewis, *Effects*, 61.

21. Rich, *Fur Trade*, 102–3.

22. Ray, *Indians*; Ray and Freeman, *"Give Us."*

23. See Trigger, "Early Native." The Blackfoot example is discussed in Binnema, "Old Swan," 12.

24. Wood and Thiessen, *Early Fur Trade*, 26, 70; Jackson, "Brandon House," 12–14.

25. B.239/a/69, 5 December 1772.

26. B. 239/a/63, 1 April 1770, B.239/a/69, 16 December 1772; Rich, *Cumberland House* (1951), Cumberland House Journals, 7 February 1777, 22 August 1779; Bain, *Travels*, 303–4; Glover, *David Thompson's Narrative*, 49.

27. Considerable and convincing evidence has now been presented that Cree bands occupied the western forests as far west as Lac la Biche well before Euroamerican traders arrived; Russell, *Eighteenth-Century Western Cree*; Smith, "Western Woods Cree." There is little evidence for the existence of the Plains Crees before the Euroamericans arrived, however.

28. Glover, *David Thompson's Narrative*, 49.

29. Rich, *Cumberland House* (1952), 11 January 1780 (p. 84).

30. John Milloy's argument that Cree bands did migrate westward in response to the trading opportunities and that the appearance of the horse may have prompted specific Cree bands to adopt a plains lifestyle is essentially correct, even if his assumption that the Crees were limited to the region east of Lake Winnipeg until Euroamericans arrived is probably incorrect. See Milloy, *Plains Cree*, 5, 24.

31. See Ray, "Northern Great Plains."

32. This evidently had been the case in the French era as it was now in the British era.

33. Rich, *Cumberland House* (1951), 111–12, 188. The Governor and Committee also sought expansion toward the plains "in hopes that We shall thereby get into possession of Wolves which are extremely valuable in Our Trade"; A. 6/11 Governor and Committee to Humphrey Marten and Council at York Factory, 14 May 1777 (fol. 104d). Also see A. 6/11, Governor and Committee to Humphrey Marten and Council at York Factory, 4 May 1780 (fol. 163).

34. Rich, *Cumberland House* (1951), lxxxiii.

35. Rich, *Cumberland House* (1951), 10 March 1776; Rich, *Cumberland House* (1952), Hudson House, 31 October 1780. Duncan McGillivray mentioned the same motivations in 1794; Morton, *Journal*, 33.

36. Simmons, "New Mexico's Smallpox," 323; Cooper, *Epidemic Disease*, 56–69. For a discussion of the epidemic on the northern plains, see Decker, "Tracing."

37. Hanson, "Spain," 11.

38. Flores, "Bison Ecology," 465.

39. Ray, *Indians*, 105–7. Ray notes that smallpox spread to the northern plains via the Dakotas of the Upper Mississippi River valley, but not the evidence that it also spread to the northwestern plains via Shoshonis.

40. B.239/a/79, fol. 73d.

41. Tyrrell, *David Thompson's Narrative*, 336–37.

42. Rich, *Cumberland House* (1952), Hudson House Journals, 22 October 1781; Cumberland House Journals, 11 December 1781. David Thompson, relying on accounts by Orcadian Mitchell Oman, suggests that Oman brought the news to Hudson House after he had been upriver near the Eagle Hills, where he had seen Indians recovering from the disease; Tyrrell, *David Thompson's Narrative*, 320–21. The Hudson House Journals, however, indicate that Oman left Hudson House on 15 October and did not return until 2 December.

43. Wood and Thiessen, *Early Fur Trade*, 71. Wood and Thiessen estimate the 1804 population of the villages at Knife River at about 3,750.

44. Rich, *Cumberland House* (1952), William Walker (Hudson House) to William Tomison (Cumberland House), 14 May 1782, 11 January 1782.

45. B.87/a/6, 4 January 1783.

46. Rich, *Cumberland House* (1952), 21 January 1782, 25 January 1782.

47. For example, see Brunton, "Smallpox Inoculation," 410–11.

48. Rich, *Cumberland House* (1952), Cumberland House Journals, 2 February 1782 (p. 234).

49. Hopkins, *Princes*, 41; Crosby, *Ecological Imperialism*, 199.

50. Brunton, "Smallpox Inoculation," 404; Crosby, *Ecological Imperialism*, 199–200; McNeill, *Plagues*, 250.

51. Hopkins, *Princes*, 72. Hopkins argues that smallpox had a tremendous influence on the course of European history.

52. Duncan, Scott, and Duncan, "Hypothesis"; Duncan, Scott, and Duncan, "Smallpox Epidemics."

53. Nicks, "Orkneymen."

54. Flinn, ed., *Scottish Population*, 291–92.

55. Brunton, "Smallpox Inoculation," 417, 412.

56. Hopkins, *Princes*, 46. Hopkins explains (p. 47) that the use of inoculation as a folk practice among illiterate people actually led to initial resistance to the practice among medical practitioners.

57. Brunton, "Smallpox Inoculation," 414.

58. Wood and Thiessen, *Early Fur Trade*, 206.

59. A discussion of this is central to Russell, *Eighteenth-Century Western Cree*.

60. Martin, *Keepers*.

61. Crosby, "Infectious Disease," 130–31.

62. William Walker noted that "the most part of them that has recover'd is women and children"; Rich, *Cumberland House* (1952), 279. Also see Tyrrell, *David Thompson's Narrative*, 323.

63. Tyrrell, *David Thompson's Narrative*, 337–38.

64. Tyrrell, *David Thompson's Narrative*, 324. Peter Fidler also noted seeing "Nee tuck kis" in 1792; E.3/2, "Journal of a Journey," 30 November 1792.

65. Glover, *David Thompson's Narrative,* 47. Young Man said that many Piegans believed that the good spirit had forsaken them and the evil spirit had destroyed them; Glover, *David Thompson's Narrative,* 49.

66. Bradley, "Journal," 166. Note that this account supports the 1805 Larocque version that the two Crow bands amalgamated after the epidemic.

67. Tyrrell, *David Thompson's Narrative,* 323. Also see B.87/a/6, William Tomison to George Hudson, 9 February 1783; B.49/a/14, 16 March 1784.

68. Tyrrell, *David Thompson's Narrative,* 338.

69. Gough, *Journal,* 522.

70. Schaeffer, "Plains Kutenai," 9.

71. B.87/a/6, 21 October 1782.

CHAPTER 7. "MANY BROILS AND ANIMOSITIES," 1782–95

1. Scholarly literature related to the topic of this chapter includes portions of Ewers, *Blackfeet: Raiders*; Fowler, *Shared Symbols*; and Milloy, *Plains Cree.* None of those works, however, deal with this period in detail.

2. Wallace, *Documents,* 7.

3. B.87/a/7, 7 August 1785; B.49/a/16, fol. 22; B.121/a/1, 2 November 1786; B.87/a/9, 17 August 1786; B.87/a/9, William Tomison to William Walker, 14 November 1786. Mackay must have hoped that the HBC presence would reduce the chance of his Canadian rivals' establishing there. Canadian traders did not fear HBC competition in these years.

4. Johnson, *Saskatchewan River,* 11n.

5. B.87/a/9, William Tomison to William Walker, 14 November 1786. Mention of Umfreville's House is in B.121/a/1.

6. Wallace, *Documents,* 16.

7. B.121/a/4, 22 October 1789; B.121/a/4, William Walker to Mitchell Oman, 11 December 1789. The HBC was apparently unable to respond because of a shortage of Brazil tobacco. In 1791 the HBC did not build higher up because many of its horses had been stolen; B.121/a/6, William Tomison to William Walker, 22 February 1792.

8. B.121/a/8, 5 June and 7 October 1792.

9. Rich, *Cumberland House* (1952), 11 November 1779, Hudson House, 28 February 1780; B.87/a/6, 1 December 1782. Even in 1785 Robert Longmoor complained that none of his men at Hudson House understood Blackfoot; B.87/a/7, 25 June 1785.

10. B.49/a/15, 15 May 1784 (fol. 65d); emphasis in the original. Perhaps Gaddy was hosted by Young Man and his band. B.121/a/1, 21 October 1786; B.121/a/2, 9 October 1787.

11. Morton, *Journal,* 41.

12. E. 3/2, "Journal of a Journey," 27 November 1792. Grace Morgan has argued that northern plains Indians had an ancient aversion to beaver hunting.

See her "Beaver Ecology." Her discussion of the importance of the place of beavers in Blackfoot mythology (pp. 34–35) and the importance of beavers in maintaining surface water sources (pp. 59–61) is intriguing. The HBCA journals, however, show that the Piegans did kill many beavers, but that their relatively sedentary lifestyle in winter precluded large-scale beaver hunting.

13. B.87/a/6, 1 and 3 March 1783 (quotation); B.87/a/6, 23 March 1783. This envoy was James Tate; B.87/a/9, William Tomison to William Walker, 14 November 1786.

14. Umfreville, *Present State,* 198.

15. B.87/a/7, 7 August 1785; B.49/a/16, William Tomison to William Walker, 11 January 1786 (fol. 22); B.121/a/1, 2 November 1786; B.121/a/2, William Tomison to William Walker, 26 December 1787 (fol. 68d).

16. B.121/a/2, 21 November 1787 and 25 March 1788. For a similar incident, see B.87/a/8, 13 April 1786.

17. E. 3/2, "Journal of a Journey," 12 December 1792, 30 and 31 December 1792 (quotations), and 1 January 1793.

18. Morton, *Journal,* 56.

19. Williams, *Andrew Graham's Observations,* 257; Morton, *Journal,* 30.

20. Wood and Thiessen, *Early Fur Trade,* 26, 70.

21. E. 3/2, "Journal of a Journey," 14 February 1793.

22. Umfreville, *Present State,* 178.

23. Swagerty, "Indian Trade," 355.

24. Kavanagh, *Comanche,* 185; Thomas Kavanagh, personal communication, 7 February 1997.

25. Umfreville, *Present State,* 188–89.

26. E. 3/2, "Journal of a Journey," 28 February 1793 and 31 December 1792.

27. Glover, *David Thompson's Narrative,* 45. Also see p. 51.

28. Tyrrell, *David Thompson's Narrative,* 338.

29. B.87/a/8, 13 April 1786; B.121/a/1, 11 March 1787 and 25 March 1788; Umfreville, *Present State,* 177; B.121/a/2, 10 February 1788.

30. E.3/2, "Journal of a Journey," 21 February 1793.

31. Glover, *David Thompson's Narrative,* 50. The Blackfoot bands had many horses with Spanish brands, bridles, and saddles in these years, but they probably did not meet the Spanish directly. A raiding party needed more than two months for a return trip to New Mexico. Contrary to Thompson's assumption, the "Black People" of this passage were probably Utes. Several plains groups called the Utes "Black People"; Hodge, *Handbook,* 2: 876.

32. E.3/2, "Journal of a Journey," 15 and 22 November 1792, 12 December 1792.

33. All the passages quoted in this paragraph are found in E.3/2, "Journal of a Journey," 12 December 1792. The Piegans themselves apparently believed that Fidler's astronomical instruments enabled him to spy on their enemies; see E.3/2, "Journal of a Journey," 17 January 1793.

34. E.3/2, "Journal of a Journey," 31 January 1793. The Missouri River was only a ten-day walk (or about five days on horseback) from Spitcheyee; Hopwood, *David Thompson*, 224.

35. E.3/2, "Journal of a Journey," 14 February 1793, 2 March 1793, and 29 and 30 December 1792.

36. Albers, "Symbiosis," particularly p. 108. During the winter of 1775–76 Alexander Henry the Elder noted that the "Black-feet" were troublesome neighbors to the Assiniboines and that small Assiniboine bands had reason to fear the Crees; Bain, *Travels*, 303, 318. This was a time of generally cordial relations among these groups, however.

37. Albers describes this general pattern on the Great Plains in "Symbiosis."

38. Morton, *Journal*, 31. Since the traders and Cree bands had long-established relationships, the traders also relied more heavily on Plains Cree bands than on other plains bands when they sought fresh and pounded meat.

39. The change was actually ordered in 1788 (A.6/11, Governor and Committee to William Tomison and Council at York Factory, 16 May 1788), but it was impossible for the orders to be transmitted inland that autumn; A.11/117, 8 September 1789 (fol. 39); B.121/a/4, 13 and 16 March 1790; B.121/a/5, 8 December 1789 and 15 March 1790 (quotation).

40. Osborn, "Ecological Aspects," 566, 580.

41. This is convincingly shown in Osborn, "Ecological Aspects."

42. Fidler mentioned "a great number of horses" among the Piegans in E.3/2, "Journal of a Journey," 18 November 1792.

43. If, as Rinn has argued, the 0° F January isotherm represents the approximate limit of equestrian culture, Cree and Assiniboine horse herds were in a very tenuous position. Evidence suggests that their herds had to be replenished continuously through raiding; Rinn, "Acquisition," 83.

44. Osborn, "Ecological Aspects," 583.

45. Bain, *Travels*, 312, 296.

46. Rinn, "Acquisition," 78–80. Peter Fidler explained in 1793 that the Blackfoot groups were "very careful" of their horses, but "the Southern Inds pays very little attention to them—& frequently they have none"; E.3/2, "Journal of a Journey," 9 January 1793; Bain, *Travels*, 312, 296.

47. Osborn, "Ecological Aspects," 584.

48. Umfreville, *Present State*, 189.

49. Umfreville, *Present State*, 200.

50. B.87/a/8, 5 May 1786; B.121/a/2, William Tomison to William Walker, 26 December 1787 (fol. 68d); B.49/a/19; B.121/a/2, 9 July 1787 and 24 August 1787 (quotation). The following day's entry, in which James Gaddy is left word on how to find this band in the fall, shows that the band knew him. On 9 October Gaddy and David Thompson did, in fact, begin their journey that would take them to Young Man's camp. The account of this journey is contained in Glover, *David Thompson's Narrative*, 46–51.

51. B.205/a/2, William Tomison to William Walker, 16 March 1788, and William Tomison to George Hudson, 13 May 1788.

52. B.121/a/7, 28 January 1792.

53. E.3/2, "Journal of a Journey," 16 November 1792. It was clear, however, that it was unusual for a Cree band to travel as far south as the Bow River; E.3/2, "Journal of a Journey," 9 January 1793. Also see E.3/2, "Journal of a Journey," 11 November 1792.

54. Morton, *Journal*, 74–75.

55. Umfreville, *Present State*, 199.

56. B.121/a/2, 13, 16, and 17 April 1788.

57. This is a reference to a combined Cree and Sarcee trading party as late as 2 May 1792; B.121/a/7. Also see B.121/a/6, 27 January 1792. They appear to have been prolific trappers.

58. Morton, *Journal*, 40.

59. Binnema, "Old Swan," especially p. 12.

60. This is the time when traders begin to mention frequently that men were employed "attending the House" to prevent theft of trade goods while plains Indian bands were at the posts. See, for example, B.121/a/2, 28 February and 25 March 1788.

61. B.121/a/1, 9 & 10 April 1787; Umfreville, *Present State*, 201; B.121/a/3, 17 May 1789 (quotation); B.121/a/5, 12 December 1789.

62. B.121/a/7, 3 October 1791; B.205/a/6, 18 September 1791, mentions that these were Bloods; B.24/a/1, William Tomison to James Tate, 30 January 1793 (fol. 45); also see B.24/a/3, 8 July 1793; B.121/a/2, 7 January 1788; B.121/a/4, 9 December 1789; Umfreville, *Present State*, 199.

63. These two maps and three other maps drawn by Siksika men are discussed in Binnema, "Indian Maps."

64. The Gros Ventre bands and traders generally communicated in Blackfoot; Umfreville, *Present State*, 198.

65. B.121/a/2, 1 May 1788, 26 April 1788. For evidence of the cool spring, see B.121/a/2, fol. 61d, which says that the North Saskatchewan River at Manchester House did not break up until 28 April.

66. B.121/a/2, 2 May 1788; B.205/a/3, 24 July 1788; B.121/a/3, 11 September 1788; B.205/a/3, William Walker to William Tomison, 12 September 1788.

67. B.121/a/3, 23 October and 29 November 1788; William Tomison to William Walker, 25 January 1789 (fol. 59d; quotation); 22 January and 16 February 1789. The Piegans did not come in until spring because of the snow; 25 April 1789. In late April two Indians visited South Branch House to borrow a horse, "their own dying through the badness of the winter"; B.205/a/3, 24 April 1789.

68. B.121/a/3, 16 and 24 May 1789. Also see B.121/a/4, 6 June 1789.

69. B.121/a/4, 11 May 1790, 24 May 1790. This is two or three weeks later than it usually broke up at Manchester House.

70. B.121/a/4, 12 May 1790.

71. For a discussion of the El Niño of 1791, see Grove, "East India Company," 125, 135, 141–42.

72. B.24/a/2, 22 October 1793; Morton, *Journal*, 62–63; B.205/a/8, 14 March 1794.

73. Morton, *Journal*, 63.

74. Morton, *Journal*, 63.

75. B.24/a/2, 8 July 1793, 11 and 12 October 1793.

76. Johnson, *Saskatchewan River*, xxxiv.

77. B.205/a/8, James Bird to William Tomison, fol. 30d.

78. The quoted passage is from B.135/a/82, fol. 57. The above account of the attack is summarized from the following: B.24/a/2, 22 October 1793; B.49/a/25a, 28 December 1793; B.205/a/8, William Tomison to James Bird, 25 October 1793 (fol. 31); F.3/1, Duncan McGillivray to Simon McTavish, 26 July 1794; Morton, *Journal*, xlix (John McDonald of Garth's version); and B.135/a/82, George Sutherland to John Thomas, 18 February 1795 (fol. 57).

79. B.205/a/8, James Bird to William Tomison, 8 November 1793, and William Tomison to James Bird, 25 October 1793 (fol. 31).

80. B.205/a/8, William Tomison to James Bird, 27 January 1794, (fols. 35d–36).

81. F.3/1, Duncan McGillivray to Simon McTavish, 26 July 1794; emphasis in the original.

82. B.205/a/8, William Tomison to James Bird, 27 January 1794, (fols. 35d–36). McGillivray's description of this incident is found in F.3/1, Duncan McGillivray to Simon McTavish, 26 July 1794. Morton, *Journal*, 44–45, reveals that this was Big Man's band. The evidence that Big Man and Gros Blanc are the same person is explained in Binnema, "Conflict?" 30–31.

83. B.24/a/2, 18 January 1794 and 5 February 1794.

84. B.135/a/82, George Sutherland to John Thomas, 18 February 1795, fol. 57d.

85. A.11/117, "YF J.C. Van Driel's Narrative," fol. 164d.

86. The attack on the South Branch Houses was remembered as a Gros Ventre attack, but some early accounts note that Siksikas were involved as well. See B.49/a/26, Magnas Twatt to Malcolm Ross, 5 September 1794.

87. B.135/a/82, George Sutherland to John Thomas, 18 February 1795, fol. 57d. Van Driel survived the attack and the fire and escaped downriver in the evening.

88. Morton, *Journal*, 14. McGillivray suggests that the NWC post was attacked before the HBC post, but he was not at South Branch House at the time. Van Driel's reports make it clear that the HBC post was attacked first. There is no shortage of accounts of this incident. The events are retold in journals by traders who were not even in the country at the time, indicating the importance of the event in the oral traditions of the traders. The account above is a summary of evidence available in the following: B.135/a/82, George Sutherland to John

Thomas, 18 February 1795, fols. 57–58; Johnson, *Saskatchewan River,* 75–76; Morton, *Journal,* 13–15; B.49/a/26, 9 July 1794, and Magnas Twatt to Malcolm Ross, 5 September 1794; A.11/117, "YF J.C. Van Driel's Narrative," fols. 163–65; F.3/1, Cornelius Van Driel to John Fisk, 18 September 1794 (fol. 195).

89. B.49/a/27a, George Sutherland to James Bird, 16 November 1795 (also in B.27.a/1, 16 November 1795). A musquetoon was a short, large-bore blunderbuss. Carlton House was relocated several times over the next decades. Curiously, it is linked with two significant events in the history of the northern plains. Constructed in response to the Gros Ventre attack of 1794, it was destroyed on 27 March 1885 during the North-West Rebellion.

90. Morton, *Journal,* 29, 17, 4.

91. Morton, *Journal,* 39, 69. They withdrew to the "Rocky Mountains," evidently meaning the Bighorn Mountains, which the Gros Ventre and Blackfoot bands considered to be part of the Rocky Mountains; see Morton, *Journal,* 27; and the Gros Ventre and Blackfoot maps in this book (figs. 7.2 to 7.5)

92. The winter of 1794–95 was very severe and snowy after the beginning of January. This clearly affected the horse herds. The Bloods and Siksikas brought their spring trade in on dogs rather than horses because the horses were in such bad condition; Morton, *Journal,* 69.

CHAPTER 8. THE APOGEE OF THE NORTHERN COALITION, 1794–1806

1. This phrase, of course, is derived from the title of Ewers's classic history, *Blackfeet: Raiders.* Other important studies that touch on the history of the northwestern plains between 1794 and 1806 include Fowler, *Shared Symbols;* and Milloy, *Plains Cree.*

2. Fleming, "Origin," 141; Wallace, *Documents,* 16. Also see Morton, *Journal,* 48, 54, 59.

3. Wallace, *Documents,* 16, 17.

4. Fleming, "Origin," 142–43; Wallace, *Documents,* 17.

5. NWC trader Daniel Williams Harmon reflected on the enmity and bloodshed in his journal entry of 28 December 1803; Lamb, *Sixteen Years.*

6. W. McGillivray, NAC MG 19 B 4, section 12.

7. Wallace, *Documents,* 19.

8. Morton, *Journal,* A-18–21.

9. Williams, "Hudson's Bay Company," 40, 43.

10. Rich, *History,* 2: 221; Williams, "Hudson's Bay Company," 42–43.

11. Morton, *Journal,* 11 May 1795.

12. Johnson, *Saskatchewan River,* 3 November 1795. These would have been established by a Mr. King on behalf of David and Peter Grant and by François Beaubien or one of his representatives; Johnson, *Saskatchewan River,* xxxii. Their share of the trade, however, was dwarfed by that of the HBC and NWC; Johnson, *Saskatchewan River,* xxxvii.

13. A.6/16, Governor and Committee to William Tomison and Council at York Factory, 31 May 1799 (fol. 78d).

14. Given the HBC's own serious labor shortage at this time, the proliferation of posts also stretched the human resources of that company significantly; Johnson, *Saskatchewan River,* 224.

15. In 1799 the companies established posts at Nelson House along the North Saskatchewan River west of Edmonton for the convenience of the Swampy Ground Assiniboines of that region and at Somerset House (at Turtlelake River not far upstream from the Eagle Hills) for the convenience of valued Cree bands of that area; Johnson, *Saskatchewan River,* 2 November 1799. Both of these posts were short-lived. Somerset House operated only for the 1799–1800 season, Nelson House for two seasons.

16. Gough, *Journal,* 519.

17. Johnson, *Saskatchewan River,* lxxi, lxxvii, lxxxi.

18. Fairfield, "Chesterfield House," 10.

19. McDonald of Garth, "Autobiographical Notes," 29.

20. Johnson, *Saskatchewan River,* 298n; Fairfield, "Chesterfield House," 38; McDonald, "Autobiographical Notes," 30. Details and debates about the operation of Chesterfield House are discussed in Binnema, "Common and Contested Ground," 299–301.

21. For a discussion of the debate over the role of Chesterfield House, see Binnema "Common and Contested Ground," 301–2; and Binnema, "Conflict?" 58–60.

22. Johnson, *Saskatchewan River,* lxxxiv.

23. Morton, *Journal,* 56.

24. Johnson, *Saskatchewan River,* 12 (quotation), 15, and 17 March 1798.

25. Johnson, *Saskatchewan River,* 26 September 1799.

26. B.60/a/5, letter of Peter Pruden (Buckingham House) to James Bird (on his way to Edmonton House), 10 August 1799.

27. Johnson, *Saskatchewan River,* 26 September, 2 November, and 4 October 1799; McDonald, "Autobiographical Notes," 26. McDonald gave the wrong year for the establishment of Rocky Mountain House.

28. Hopwood, *David Thompson,* 216, 218. Fidler believed that this was the first visit to a trading post that these Kutenais had ever made; E.3/2, "Journal of a Journey," fols. 19d–20. This entry is dated 31 December 1792 but was obviously written later.

29. Hopwood, *David Thompson,* 223.

30. B.60/a/6, James Bird to John McNab, 23 December 1806 (fol. 6); 30 April 1807; and James Bird to John Peter Pruden, 2 May 1807.

31. Fidler believed that no Euroamericans had ever traveled on the South Saskatchewan River above the site of the abandoned South Branch House (Johnson, *Saskatchewan River,* 15 August 1800), but one or more Euroamericans may have done so in the 1750s or 1760s.

32. Binnema, "Conflict?" 65–69.

33. Johnson, *Saskatchewan River,* 21 August 1800, 3 September 1800, and 28 March 1801.

34. Only a few months earlier a few Canadian traders heading down the North Saskatchewan River (who were apparently initially mistaken for Indians) were attacked, wounded, and captured by some Gros Ventres, who took several of their guns before releasing them; Johnson, *Saskatchewan River,* 24 and 25 May 1800. None of the wounds were very serious, but this event would have refreshed the traders' memories of the Gros Ventre attacks of the 1790s.

35. Johnson, *Saskatchewan River,* 31 October 1801. Elsewhere Fidler wrote that these Arapahos had come from "the Red Deers River [Yellowstone] beyond the Big River [Missouri]" (B.39/a/2, 30 October 1801), but a 44-day journey suggests that they had come from farther than that.

36. A fine explication of this process is given in J. Foster, "Wintering," 1–7.

37. E.3/2, "Journal of a Journey," 18 March 1793.

38. Johnson, *Saskatchewan River,* xci–xcii; Tyrrell, *David Thompson's Narrative,* 312.

39. Hopwood, *David Thompson,* 223–24.

40. Tyrrell, *David Thompson's Narrative,* 312–17. In this passage Thompson relates how a band of Gros Ventres killed twenty-five Iroquois near the foothills of the Rockies, probably in 1801. Thompson also suggests that the local Crees were ambivalent toward the newcomers. The Iroquois in the west are discussed in some detail in Nicks, "Orkneymen," 85–101.

41. E.3/2, "Journal of a Journey," 29 December 1792. Fidler here gauged Young Man to be the second most prominent Piegan leader, an assessment that accords with David Thompson's judgment.

42. Tyrrell, *David Thompson's Narrative,* 346; E.3/2, "Journal of a Journey," 9 November 1792.

43. Tyrrell, *David Thompson's Narrative,* 347.

44. Morton, *Journal,* 50.

45. Binnema, "Old Swan."

46. Morton, *Journal,* 50.

47. Johnson, *Saskatchewan River,* 24 October 1801.

48. Morton, *Journal,* 45–46. Sitting Badger is discussed on pp. 75–76.

49. Morton, *Journal,* 44; Gough, *Journal,* 381, 424, 542, 546.

50. This interpretation is based on evidence that traders had known this younger man as "Feathers" or "Painted Feathers" and continued to refer to him by that name after 1794, probably to avoid confusing him with his father. His Blackfoot name was given as "Akkomakki," however, which translates as "Old Swan," not "Feathers"; B.39/a/2, fol. 98d. For a survey of the evidence of Old Swan's life, see Binnema, "Old Swan."

51. Old Swan evidently pursued a confrontational policy toward Euroamerican traders before 1794; see Morton, *Journal,* 50; and Binnema, "Old Swan," 20–21.

52. Lamb, *Sixteen Years,* 57, 213; Glover, *David Thompson's Narrative,* 267; Johnson, *Saskatchewan River,* 65, 67, 142; Gough, *Journal,* 396.

53. On 18 August 1803 Harmon mentioned that Crees and Assiniboines of the Swan River district were trying to make peace with the Gros Ventres and Siksikas after protracted warfare; Lamb, *Sixteen Years,* 69. The evidence from the North Saskatchewan River journals would suggest that the Gros Ventres were the primary targets of this warfare, but Harmon's reference to the Siksikas must refer to Big Man's bands.

54. Johnson, *Saskatchewan River,* 21 February 1800, 19 February 1800, 118 and 133.

55. E.3/2, "Journal of a Journey," 14 March 1793.

56. Johnson, *Saskatchewan River,* 12 February 1798.

57. Since Napi was neither omnipotent nor always benevolent, the Blackfoot people were ambivalent about him. See Binnema, "Old Swan," 12. Documentary evidence about initial contacts between Euroamericans and Indians of the northwestern plains is scant; thus firm conclusions are impossible. For a stimulating discussion of initial contacts in other regions, see Trigger, "Early Native."

58. Morton, *Journal,* 50.

59. Lamb, *Sixteen Years,* 73.

60. Regarding the price of wolf furs, see A.6/16, Governor and Committee to William Tomison and Council at York Factory, 31 May 1799 (fol. 78). Regarding the high duty (7 pence per skin) and the order to reduce the trading standard, see A.6/16, Governor and Committee to John Ballenden and Council at York Factory, 28 May 1800 (fol. 104). The general recovery in fur prices over the next few years did not include wolf; A.6/17, Governor and Committee to John McNab and Council at York Factory, 31 May 1806 (fol. 77d).

61. Johnson, *Saskatchewan River,* lxxix.

62. Johnson, *Saskatchewan River,* 163. Even as late as 1797 James Gaddy Sr. was described as "the only One [in the HBC] that Speaks the Blackfoot & Blood Indian Language"; Johnson, *Saskatchewan River,* 9n.

63. Johnson, *Saskatchewan River,* 23 January 1796.

64. Lamb, *Sixteen Years,* 30 July 1803.

65. See Binnema, "Old Swan," 16–20.

66. For more detail, see Binnema, "Old Swan."

67. Johnson, *Saskatchewan River,* 8 January 1802.

68. Gough, *Journal,* 378–97, 540.

69. B.34/a/2, December 19, 1800.

70. Johnson, *Saskatchewan River,* 1 November 1800, p. 278 (quotation), 27 December 1800, 27 February 1801, 8 January 1802, 25 January 1802.

71. Some have been led to assume that these "Mud House Indians" were Hidatsas and that Old Swan misplaced the village on the map; Moodie and Kaye, "Ac Ko Mok Ki Map," 6. In 1802, however, Charles Le Raye encountered Crow winter camps that were "sunk three feet below the surface of the ground, but

otherwise are built nearly similar to those of the Gros Ventres [Hidatsas]"; Le Raye, "Journal," 172. Mud house villages were clearly still to be found well above the Hidatsa villages. Old Swan probably did not misplace the village on the map but may have been noting the site of a Crow (or Hidatsa) winter camp. On their trips up and down the Missouri River in 1805 and 1806 Lewis and Clark saw abandoned camps of various descriptions (although they do not mention mud houses) almost daily; Moulton, *Journals.* They saw two abandoned encampments (including one of 126 lodges) at the confluence of the Missouri and Judith Rivers, suggesting that this location was an important Indian campsite; Moulton, *Journals,* 29 May 1805 (4: 216, 219).

72. The Blackfoot bands clearly dominated the fescue belt well into the Missouri basin. Meriwether Lewis, aside from his meeting with the Piegans in the Marias basin in 1806, also saw ample evidence of Piegan winter camps and summer camps near that river; Moulton, *Journals,* 25 July 1806 (8: 127).

73. Cooper found that Mexico City was hit by smallpox in 1797–98; Cooper, *Epidemic Disease,* 86–156; Johnson, *Saskatchewan River,* 294.

74. Wood and Thiessen, *Early Fur Trade,* 206. They had 600 warriors after the epidemic. Larocque explained that the Crows had numbered 2,000 lodges, but this was apparently before the epidemic of the 1780s. The mortality of the 1801 epidemic is not clearly given.

75. Teit, "Salishan Tribes," 315.

76. Not surprisingly, members of different ethnic groups could be found camping together. For example, in 1805 some Shoshonis lived with a Crow band, apparently on a long-term basis; Wood and Thiessen, *Early Fur Trade,* 170, 220.

77. Le Raye, "Journal," 173.

78. Wood and Thiessen, *Early Fur Trade,* 220.

79. Wood and Thiessen, *Early Fur Trade,* 219. The Flatheads almost certainly also would have traded products like dentalium shells that they acquired from their western neighbors at the time.

80. First quotation from Wood and Thiessen, *Early Fur Trade,* 220 (emphasis in original); second quotation from Gough, *Journal,* 26 July 1806. Also see Moulton, *Journals,* 14 August 1805 (5: 91).

81. Wood and Thiessen, *Early Fur Trade,* 217, 220; Moulton, *Journals,* 17 August 1805 (5: 114). The Shoshonis apparently valued blue beads highly. They demanded a horse from Larocque in exchange for a hundred of them; Wood and Thiessen, *Early Fur Trade,* 192.

82. The Mandan and Hidatsa villages, which had been consolidated at Knife River after the smallpox of the 1780s, probably had fewer than 4,000 people in 1800; Wood, *Papers,* 18.

83. Wood and Thiessen, *Early Fur Trade,* 213. While Larocque was apparently impressed by the profits of the Crows, Alexander Henry the Younger, when witnessing the trade at the Hidatsa villages, remarked that "it was really disgusting to see how those impious vagabonds the Big Bellies kept those poor inoffensive

Crow Indians in subjection, during their stay at the Villages, making their own price for their horses, and every other they bring, nor will they allow a stranger to give the real value of their commodities"; Gough, *Journal*, 27 July 1806. Henry, of course, was considering that the Hidatsas themselves resold their horses at considerably higher prices. For a discussion of Hidatsa horse trading practices in these years, see Wood and Thiessen, *Early Fur Trade*, 65–67.

84. Wood and Thiessen, *Early Fur Trade*, 215, 170, 246; Gough, *Journal*, 26 and 27 July 1806.

85. Wood and Thiessen, *Early Fur Trade*, 170.

86. Compare Abel, *Tabeau's Narrative*, 160–61, with Old Swan's map of 1801 (figure 7.2).

87. Wood and Thiessen, *Early Fur Trade*, 213, 245.

88. Le Raye, "Journal," 170, 175.

89. Larocque quoted in Wood and Thiessen, *Early Fur Trade*, 213 (quotation), 245, 206. The lack of references in the HBC documents to Crow raids on the northern coalition at this time suggests that Larocque was correct.

90. McKenzie quoted in Wood and Thiessen, *Early Fur Trade*, 245, 248.

91. Wood and Thiessen, *Early Fur Trade*, 219–20.

92. Le Raye, "Journal," 13 October 1802. Patrick Gass, a member of the Lewis and Clark expedition, concurred with the common assessments of the poverty of the Shoshonis but attributed the attacks upon them to different motives. He found the Shoshonis "the poorest and most miserable nation I ever beheld; having scarcely any thing to subsist on, except berries and a few fish, which they contrive by some means, to take. They have a great many fine horses, and nothing more; and on account of these they are much harassed by other nations"; Moulton, *Journal of Patrick Gass*, vol. 10 of *Journals*, 20 August 1805 (10: 128).

93. Le Raye, "Journal," 174. Lewis and Clark similarly described the arms of the Shoshonis as consisting of bows and arrows, shields, lances, and "pog-gar'-mag-gon'" (war clubs); Moulton, *Journals*, 19 August 1805 (5: 122).

94. Johnson, *Saskatchewan River*, 25 January 1802; B.39/a/2, 25 January 1802. In the summer of 1805 Lewis and Clark saw a few NWC guns among Shoshoni bands. The Shoshonis had acquired them from the Crows; Moulton, *Journals*, 13 August 1805 (5: 80).

95. Moulton, *Journals*, 19 August 1805 (5: 123). The reference to "extreem poverty" is in Moulton, *Journals*, 19 August 1805 (5: 119).

96. The Siksikas killed four of the five Shoshonis in the party; Johnson, *Saskatchewan River*, 8 and 25 January 1802.

97. Wood and Thiessen, *Early Fur Trade*, 218–19. Moulton, *Journals*, 14 August 1805 (5: 91). Meriwether Lewis, however, reported that bison could be found very near the Great Divide in the upper Missouri basin; Moulton, *Journals*, 7 July 1806 (8: 96).

98. Bird Woman (Sakakewea or Sakajewea), Lewis and Clark's famous Shoshoni guide, told William Clark that only a few years earlier the bison had

been abundant well above the Three Forks of the Missouri; Moulton, *Journals,* 14 July 1806 (8: 182). By 1805 and 1806, however, they were rare there, despite the fine grasses that these valleys supported; Moulton, *Journals,* especially 8 August 1805 (5: 58).

99. Hopwood, *David Thompson,* 224; Le Raye, "Journal," 175; Wood and Thiessen, *Early Fur Trade,* 219; Moulton, *Journals,* 18 August 1805 (5: 123).

100. Le Raye, "Journal," 175; Wood and Thiessen, *Early Fur Trade,* 219.

101. The foregoing account of the winter of 1800–1801 is summarized from E.3/2, "Journal of a Journey," fols. 19d–20.

102. For a discussion of deteriorating relations between the Cree and Assiniboine bands and the Mandan and Hidatsa villagers, see Milloy, *Plains Cree,* 47–58.

103. Wood and Thiessen, *Early Fur Trade,* 247.

104. Moulton, *Journals,* 28 July 1805 (5: 8–9).

105. Ewers, *Indian Life,* 59.

106. McDonald, "Autobiographical Notes," 31–32.

107. Wood and Thiessen, *Early Fur Trade,* 233.

108. McDonald, "Autobiographical Notes," 34. McDonald implied that two other Canadians also died in the attack. Lewis and Clark noted that the Hidatsa war party of about fifteen men had left the middle Missouri villages in March 1805; Moulton, *Journals,* 17 May 1805 (4: 159–60, 161). A description of this incident is also preserved in HBC documents, which note that Howse's party escaped harm because they had passed this point a few days earlier; B.239/a/111, 29 June 1805.

109. Wood and Thiessen, *Early Fur Trade,* 166. Larocque's report is supplemented by Charles McKenzie's on p. 233.

110. A.6/15, Governor and Committee to William Tomison and Council at York Factory, 30 May 1795, (fol. 138d). The Governor and Committee was commenting on the attack on Manchester House and apparently had not yet learned of the South Branch House incident.

111. Morton, *Journal,* 32, 44–46. For a discussion, see Binnema, "Old Swan," 16.

112. Morton, *Journal,* 73. This was apparently an about-face for Tomison, for on 28 February Duncan McGillivray expressed his surprise that Tomison "seems to entertain but little resentment for those acts of barbarity and injustice"; Morton, *Journal,* 57.

113. McDonald as quoted in Morton, *Journal,* A-3. Further discussion of the disputes that arose from this incident is found in Johnson, *Saskatchewan River,* xxiii–xxxiv.

114. Johnson, *Saskatchewan River,* 16 December 1796.

115. Morton, *Journal,* 39. McGillivray's use of the term "Slave" suggests that the expedition was aimed at more than just the Gros Ventres. Big Man's Siksika bands probably also would have been targets.

116. B.148/a/1, 7 January 1795, and letter from James Bird to William Tomison, 15 January 1795 (fol. 32).

117. Morton, *Journal,* 78.

118. Tyrrell, *David Thompson's Narrative*, 327.

119. Morton, *Journal*, 14.

120. Johnson, *Saskatchewan River*, 6 and 8 March 1802, 29 October 1801 (quotation).

121. Morton, *Journal*, 46.

122. Johnson, *Saskatchewan River*, 12 November 1800 and 1 December 1800; E.3/2, 20 September 1800.

123. Johnson, *Saskatchewan River*, 17 February 1801, 299n, 290 (quotation), 317n. The announced intentions of the Gros Ventres in Johnson, *Saskatchewan River*, 19 February and 2 March 1801, and the location of subsequent events suggest that they would have been in the upper South Saskatchewan basin at this time.

124. Johnson, *Saskatchewan River*, 27 September 1801, 297n, 1 February 1802, 294. The epidemic would have killed young people disproportionately because older members of the community would have acquired immunity from their exposure to the epidemic of 1781.

125. Rich, *History*, 2: 221; and Johnson, *Saskatchewan River* lxxix, 11 April 1801, 319n.

126. B.39/a/2, 28 October 1801, 1 March 1802 (quotations; emphasis in original), 1 February 1802, 27 January 1802 (quotation).

127. Johnson, *Saskatchewan River*, 9 February 1802, 1 February 1802.

128. Johnson, *Saskatchewan River*, 3 October 1801, 1 February 1802.

129. Johnson, *Saskatchewan River*, 21 February 1802; B.39/a/2, 22 February 1802. Feet, hands, noses, and private parts were cut off, and bellies were cut open; B.39/a/2, 25 February 1802.

130. B.39/a/2, 1 March 1802.

131. Johnson, *Saskatchewan River*, 3–4 and 6 March 1802; B.39/a/2, 3 March 1802, 4 March 1802.

132. The available documentary evidence relating to Little Bear is summarized in Binnema, "Conflict?" 65–96.

133. Johnson, *Saskatchewan River*, 15 April 1802.

134. On 16 February 1801 about a hundred Gros Ventres left Chesterfield House on a war expedition against the Shoshonis; Johnson, *Saskatchewan River*, 16 February 1801. Peter Fidler bought a Shoshoni slave girl about six years old from the Gros Ventres for 7 MB in 1801, but the Gros Ventres wanted her back the next day; B.39/a/2, 6 and 7 November 1801. On 18 February 1802 he bought a Shoshoni slave woman nineteen years old from a Gros Ventre man. This woman was also captured from the Crows in warfare. Fidler was asked to return her after one day; Johnson, *Saskatchewan River*, 311n.

135. Moulton, *Journals*, 13 August 1805 (5: 83).

136. Le Raye, "Journal," 170, 175.

137. Wood and Thiessen, *Early Fur Trade*, 191. John McDonald of Garth was known as "bras croche" among voyagers because of his crippled right arm; Morton, *Journal*, xliv.

138. Lamb, *Sixteen Years,* 18 August 1803 (quotation). Harmon had discovered the campsite of a recently abandoned Gros Ventre war party less than a month before; Lamb, *Sixteen Years,* 30 July 1803, 19 (quotation) and 28 April 1806. The NWC had reestablished South Branch House in the summer of 1804; Lamb, *Sixteen Years,* 21 September 1805. From 1805 to 1810 the HBC's Carlton House was also located at the site of old South Branch House.

EPILOGUE

1. See Binnema, "Old Swan."

2. Lamb, *Sixteen Years,* 100.

3. B.60/a/6, James Bird to John McNab, 23 December 1806 (fol. 5; quotation); 25 August 1806; 22 September 1806; B.60/a/6, James Bird to John McNab, 23 December 1806 (fol. 5; quotation).

4. In *Plains Cree* Milloy has summarized Cree-Blackfoot relations from the 1790s to 1870. For his explanation of 1806 as a turning point, see his chapter 3.

5. Hopwood, *David Thompson,* 243.

BIBLIOGRAPHY

ARCHIVES AND MANUSCRIPTS

Glenbow Alberta Institute (GAI), Calgary

M 736: "Description of a Journey from Fort Benton to Bow River, NWT." Duncan McNab McEachran fonds.

M 4376: David C. Duvall fonds.

M 4421: Robert Nathaniel Wilson fonds.

M 4436: Claude Everett Schaeffer fonds.

M 8462: Oscar Lewis fonds.

Hudson's Bay Company Archives (HBCA), Winnipeg

Headquarter Records

A.6/11–: London Correspondence Book Outwards, 1770–.

A.11/114–: London Correspondence Inwards from HBC Posts, York Factory, 1716–.

Post Journals and Records

B.24/a/1–: Buckingham House Post Journals, 1792–99.

B.27/a/1–: Carlton House Post Journals (Saskatchewan), 1795–.

B.28/a/4: Carlton House Post Journals (Assiniboine), 1797.

B.34/a/1–3: Chesterfield House Post Journals, 1801–2.

B.39/a/2: Chesterfield House Rough Journals, 1801–2.

B.49/a/1–35: Cumberland House Post Journals, 1774–1804.

B.49/a/14–15: Hudson House Journals 1783–84 and 1784–85.

B.49/a/27[b]: Buckingham House Rough Journals, 1796–97.

B.60/a/1–: Edmonton House Post Journals.

B.60/z/1–2: Edmonton House Miscellaneous, 1810–1932.

B.87/a/1–9: Hudson House Post Journals, 1778–87.

B.92/a/1: Island House Post Journals, 1800–1801.

B.121/a/1–8: Manchester House Post Journals, 1786–93.

B.135/a/82: Moose Fort Journals, 1794–95.
B.148/a/1: Nipawi Post Journals, 1794–95.
B.204/a/1: Somerset House (Turtle Creek), 1799–1800.
B.205/a/1–8: South Branch Post Journals, 1786–94.
B.239/a/1–105: York Factory Post Journals, 1714–1801.

Private Papers

E.3/2, fols. 2–39: "Journal of a Journey over Land from Buckingham House to the Rocky Mountains in 1792 & 3 by Peter Fidler."
E.3/2, fols. 62d–72: "Journal from the mouth of the South Branch of the Saskatchewan River to the confluence of the Bad & Red Deers Rivers where Chesterfield House is situated by Peter Fidler, 1800."

North West Company Documents

F.3/11: North West Company Correspondence, etc., 1791–99.

National Archives of Canada (NAC), Ottawa

MG 19 B 4: William McGillivray, "Sketch of the Fur Trade, 1809."

BOOKS AND ARTICLES

Abel, Annie Heloise, ed. *Tabeau's Narrative of Loisel's Expedition to the Upper Missouri.* Norman: University of Oklahoma Press, 1939.

Albers, Patricia C. "Changing Patterns of Ethnicity in the Northeastern Plains, 1780–1870." In Jonathan D. Hill, ed., *History, Power, and Identity: Ethnogenesis in the Americas, 1492–1992,* 90–118. Iowa City: University of Iowa Press, 1996.

———. "Symbiosis, Merger, and War: Contrasting Forms of Intertribal Relationship among Historic Plains Indians." In John H. Moore, *The Political Economy of North American Indians,* 94–132. Norman: University of Oklahoma Press, 1993.

Alwin, John A. "Pelts, Provisions, and Perceptions: The Hudson's Bay Company Mandan Indian Trade, 1795–1812." *Montana* 29 (July 1979): 16–27.

Archibold, O. W., and M. R. Wilson. "The Natural Vegetation of Saskatchewan Prior to Agricultural Settlement." *Canadian Journal of Botany* 58 (1980): 2031–42.

Arthearn, Frederic J. "A Time of Transition: New Mexico in the Eighteenth Century." *Southwestern Lore* 59 (1) (1993): 16–25.

Arthur, George W. *An Archaeological Survey of the Upper Yellowstone River Drainage, Montana.* Report No. 26. Bozeman, Mont.: Agricultural Economics Research, 1966.

———. *An Introduction to the Ecology of Early Historic Communal Bison Hunting among the Northern Plains Indians.* Mercury Series Paper No. 37. Ottawa: National Museum of Man, 1975.

Arthur, George W., Michael Wilson, and Richard G. Forbis. *The Relationship of Bison to the Indians of the Great Plains.* Ottawa: National Historic Parks and Sites Branch, Parks Canada and Department of Indian Affairs, 1975.

Axelrod, Daniel J. "Rise of the Grassland Biome, Central North America." *Botanical Review* 51 (1985): 163–201.

Bailey, Arthur W. "Aspen Parkland: Then and Now." *Alberta Archaeological Review* 24 (1992): 21–23.

Bailey, Arthur W., and Robert A. Wroe. "Aspen Invasion in a Portion of the Alberta Parklands." *Journal of Range Management* 27 (1974): 263–66.

Bain, James, ed. *Travels and Adventures in Canada and the Indian Territories between the Years 1760 and 1776 by Alexander Henry, Fur Trader.* Toronto: George N. Morang, 1901.

Bamforth, Douglas B. *Ecology and Human Organization on the Great Plains.* New York: Plenum Press, 1988.

———. "Historical Documents and Bison Ecology on the Great Plains." *Plains Anthropologist* 32 (1987): 1–16.

———. "Indigenous People, Indigenous Violence: Precontact Warfare on the North American Great Plains." *Man: The Journal of the Royal Anthropological Institute* 29 (1994): 95–116.

Barrett, Stephen W., and Stephen F. Arno. "Indian Fires as an Ecological Influence in the Northern Rockies." *Journal of Forestry* 80 (1982): 647–51.

Barry, P. S. *Mystical Themes in Milk River Rock Art.* Edmonton: University of Alberta Press, 1991.

Beaty, Chester B. *The Landscapes of Southern Alberta: A Regional Geomorphology.* Lethbridge: University of Lethbridge Printing Services, 1975.

Berkhofer, Robert F., Jr. *The White Man's Indian: Images of the American Indian from Columbus to the Present.* New York: Knopf, 1978.

Bettinger, Robert L. "How, When, and Why Numic Spread." In David B. Madsen and David Rhode, eds., *Across the West: Human Population Movement and the Expansion of the Numa,* 44–55. Salt Lake City: University of Utah Press, 1994.

Bettinger, Robert L., and Martin A. Baumhoff. "The Numic Spread: Great Basin Cultures in Competition." *American Antiquity* 47 (1982): 485–503.

———. "Return Rates and Intensity of Resource Use in Numic and Prenumic Adaptive Strategies." *American Antiquity* 48 (1983): 830–34.

Binnema, Theodore. "The Common and Contested Ground: A History of the Northwestern Plains from A.D. 200 to 1806." Ph.D. dissertation, University of Alberta, 1998.

———. "Conflict or Cooperation?: Blackfoot Trade Strategies, 1794–1815." M.A. thesis, University of Alberta, 1992.

———. "Indian Maps as Ethnohistorical Sources." Unpublished paper presented at the Thirtieth Northern Great Plains History Conference, 27–30 September 1995, Brandon, Manitoba.

———. "Old Swan, Big Man, and the Siksika Bands, 1794–1815." *Canadian Historical Review* 77 (1996): 1–32.

Biolsi, Thomas. "Ecological and Cultural Factors in Plains Indian Warfare." In R. Brian Ferguson, ed., *Warfare, Culture, and Environment,* 141–68. Orlando: Academic Press, 1984.

Bird, Ralph D. *Ecology of the Aspen Parkland of Western Canada.* Publication No. 1066. Ottawa: Department of Agriculture, 1961.

Black, W. H., A. L. Baker, V. I. Clark, and O. R. Matthews. *Effect of Different Methods of Grazing on Native Vegetation and Gains of Steers in Northern Great Plains.* Technical Bulletin No. 547. Washington, D.C.: United States Department of Agriculture, 1937.

Bradley, James H. "Journal of James H. Bradley." *Contributions to the Historical Society of Montana* 2 (1896): 140–228.

———. "Lieut. James H. Bradley Manuscript." *Contributions to the Historical Society of Montana* 9 (1923): 226–351.

Brasser, R. J. "The Sarsi: Athapaskans on the Northern Plains." *Arctic Anthropology* 28 (1) (1991): 67–73.

Brink, Jack. *Dog Days in Alberta.* Occasional Paper No. 28, Alberta Culture. Edmonton: Archaeological Survey of Alberta, 1986.

Brink, Jack, and Bob Dawe. *Final Report of the 1985 and 1986 Field Season at Head-Smashed-In Buffalo Jump, Alberta Culture.* Edmonton: Archaeological Survey of Alberta, 1989.

Brumley, John H. *Ramillies: A Late Prehistoric Bison Kill and Campsite Located in Southeastern Alberta, Canada.* Mercury Series Paper No. 55. Ottawa: Archaeological Survey of Canada, 1976.

Brunton, Deborah. "Smallpox Inoculation and Demographic Trends in Eighteenth-Century Scotland." *Medical History* 36 (1992): 403–29.

Buchner, Anthony P. "The Geochronology of the Lockport Site." *Manitoba Archaeological Quarterly* 12 (2) (April 1988): 27–31.

Burpee, Lawrence J., ed. "An Adventurer from Hudson Bay: Journal of Matthew Cocking from York Factory to the Blackfeet Country, 1772–1773." *Royal Society of Canada Proceedings and Transactions,* series 3, vol. 2 (1908): 89–121.

———, ed. *Journals and Letters of Pierre Gaultier de Varrennes de la Vérendrye and His Sons.* Toronto: Champlain Society, 1927.

———, ed. "York Factory to the Blackfeet Country: The Journal of Anthony Hendry, 1754–1755." *Transactions of the Royal Society of Canada,* series 3, vol. 1 (1907), section 2: 307–61.

Butler, B. Robert. "Bison Hunting in the Desert West." In Leslie B. Davis and Michael Wilson, eds., *Bison Procurement and Utilization: A Symposium,* 106–12. Memoir 14, *Plains Anthropologist* 23, pt. 2 (1978).

Byrne, William J. "An Archaeological Demonstration of Migration on the Northern Great Plains." In Robert C. Dunnell and Edwin S. Hall Jr., eds., *Archaeological Essays in Honor of Irving B. Rouse,* 247–73. The Hague: Mouton Publishers, 1978.

———. *The Archaeology and Prehistory of Southern Alberta as Reflected by Ceramics.* Mercury Series Paper No. 14. Ottawa: Archaeological Survey of Canada, 1973.

Calloway, Colin G. "The Inter-Tribal Balance of Power on the Great Plains, 1760–1850." *Journal of American Studies* 16 (1982): 25–47.

———. "The Only Way Open to Us: The Crow Struggle for Survival in the Nineteenth Century." *North Dakota History* 53 (3) (Summer 1986): 25–34.

———. "Snake Frontiers: The Eastern Shoshones in the Eighteenth Century." *Annals of Wyoming* 63 (3) (1991): 82–92.

Campbell, Celina, Ian D. Campbell, Charles B. Blyth, and John H. McAndrews. "Bison Extirpation May Have Caused Aspen Expansion in Western Canada." *Ecography* 17 (1994): 360–62.

Chamberlain, Alexander F. "Report on the Kootenay Indians of South-Eastern British Columbia." *Report of the British Association for the Advancement of Science, Annual Meeting Report* 62 (1892): 549–617.

———. "Some Kutenai Linguistic Material." *American Anthropologist* 11 (1909): 13–26.

Chittenden, Hiram. *The History of the American Fur Trade of the Far West.* New York: Francis P. Harper, 1902.

Christopherson, R. J., and R. J. Hudson. "Effects of Temperature and Wind on Cattle and Bison." *Annual Feeders' Day Report* (28 June 1978): 40–41.

Christopherson, R. J., R. J. Hudson, and R. J. Richmond. "Feed Intake, Metabolism and Thermal Insulation of Bison, Yak, Scottish Highland and Hereford Calves during Winter." *Annual Feeders' Day Report* (8 June 1976): 51–52.

Clarke, S. E., et al. *An Ecological and Grazing Capacity Study of the Native Grass Pastures in Southern Alberta, Saskatchewan and Manitoba.* Publication No. 738. Ottawa: Department of Agriculture, 1942.

Clarke, S. E., E. W. Tisdale, and N. A. Skoglund. *The Effects of Climate and Grazing Practices on Short-Grass Prairie Vegetation in Southern Alberta and Southwestern Saskatchewan.* Publication No. 747. Ottawa: Department of Agriculture, 1947.

Clifton, James A. "The Tribal History—An Obsolete Paradigm." *American Indian Culture and Research Journal* 3 (4) (1979): 81–100.

Conner, Stuart W., and Betty Lu Conner. *Rock Art of the Montana High Plains.* Santa Barbara: University of California, Santa Barbara, 1971.

Cooper, Donald B. *Epidemic Disease in Mexico City 1761–1813: An Administrative, Social, and Medical Study.* Austin: University of Texas Press, 1965.

Coues, Elliot, ed. *New Light on the Early History of the Greater Northwest: The Manuscript Journals of Alexander Henry [the Younger], Fur Trader of the North West Company and of David Thompson, 1799–1814.* Vol. 2. Minneapolis: Ross & Haines, 1965.

Coupland, Robert T. "A Reconsideration of Grassland Classification in the Northern Great Plains of North America." *Journal of Ecology* 49 (1961): 135–67.

Coupland, Robert T., and T. Christopher Brayshaw. "The Fescue Grassland in Saskatchewan." *Ecology* 34 (1953): 386–405.

Courtney, Rick F. "Pronghorn Use of Recently Burned Mixed Prairie in Alberta." *Journal of Wildlife Management* 53 (1989): 302–5.

Crosby, Alfred W. *The Columbian Exchange: Biological and Cultural Consequences of 1492.* Westport, Conn.: Greenwood Press, 1972.

———. *Ecological Imperialism: The Biological Expansion of Europe, 900–1900.* Cambridge: Cambridge University Press, 1986.

———. "Infectious Disease and the Demography of the Atlantic Peoples." *Journal of World History* 2 (1991): 119–33.

———. "The Past and Present of Environmental History." *American Historical Review* 100 (1995): 1177–89.

———. "Virgin Soil Epidemics as a Factor in the Aboriginal Depopulation of America." *William and Mary Quarterly* 33 (1976): 289–99.

Damas, David. *Contributions to Anthropology: Band Societies.* Bulletin 228, Anthropological Series 84. Ottawa: National Museum of Canada, 1969.

Daubenmire, R. "Ecology of Fire in Grasslands." *Advances in Ecological Research* 5 (1968): 209–66.

Davis, L. B., and J. W. Fisher. "Avonlea Predation on Wintering Plains Pronghorn." In Leslie B. Davis, ed., *Avonlea Yesterday and Today: Archaeology and Prehistory,* 101–18. Saskatoon: Saskatchewan Archaeological Society, 1988.

Davis, Leslie B., ed. *Avonlea Yesterday and Today: Archaeology and Prehistory.* Saskatoon: Saskatchewan Archaeological Society, 1988.

———, ed. *Symposium on the Crow-Hidatsa Separations.* Special issue of *Archaeology in Montana* 20 (3) (September–December 1979).

Davis, Leslie B., and B. O. K. Reeves, eds. *Hunters of the Recent Past.* London: Unwin Hyman, 1990.

Davis, Leslie B., and Michael Wilson, eds. *Bison Procurement and Utilization: A Symposium.* Memoir 14, *Plains Anthropologist* 23 (pt. 2) (1978).

Dawson, George M. *Preliminary Report on the Physical and Geological Features of That Portion of the Rocky Mountains, between Latitudes 49° and 51° 30′.* Geological Survey of Canada, Annual Report, 1885. Montreal: Geological Survey of Canada, 1886.

———. *Report on the Region in the Vicinity of the Bow and Belly Rivers.* Geological Survey of Canada, Report of Progress, 1882–84, Section C. Montreal: Geological Survey of Canada, 1884.

Decker, J. F. "Tracing Historical Diffusion Patterns: The Case of the 1780–82 Smallpox Epidemic among the Indians of Western Canada." *Native Studies Review* 4 (1988): 1–24.

Dempsey, Hugh A. "The Blackfoot Indians." In R. Bruce Morrison and C. Roderick Wilson, eds., *Native Peoples: The Canadian Experience,* 404–35. Toronto: McClelland & Stewart, 1986.

Denig, Edwin Thompson. *Five Indian Tribes of the Upper Missouri.* Edited and with an introduction by John C. Ewers. Norman: University of Oklahoma Press, 1961.

Dictionary of Canadian Biography. Toronto: University of Toronto Press, 1966–.

Doige, Gary. "Warfare and Alliance Patterns of the Assiniboine to 1807." M.A. thesis, University of Manitoba, 1989.

Doughty, A., and C. Martin, eds. *The Kelsey Papers.* Ottawa: Public Archives of Canada and the Public Record Office of Northern Ireland, 1929.

Douglas, R., and J. N. Wallace, eds. *Twenty Years of York Factory, 1694–1714: Jérémie's Account of Hudson Strait and Bay.* Ottawa: Thorburn and Abbott, 1926.

Duke, Philip, and Michael Clayton Wilson. "Cultures of the Mountains and the Plains: From the Selkirk Mountains to the Bitterroot Range." In Karl H. Schlesier, ed., *Plains Indians, A.D. 500–1500: The Archaeological Past of Historic Groups,* 56–70. Norman: University of Oklahoma Press, 1994.

Duncan, Stephen R., Susan Scott, and Christopher J. Duncan. "An Hypothesis for the Periodicity of Smallpox Epidemics as Revealed by Time Series Analysis." *Journal of Theoretical Biology* 160 (1993): 231–48.

———. "Smallpox Epidemics in Cities in Britain." *Journal of Interdisciplinary History* 25 (1994): 255–71.

Eccles, W. J. *The Canadian Frontier, 1534–1760.* Toronto: Holt, Rinehart & Winston, 1969.

Emerson, Thomas E., and R. Barry Lewis, eds. *Cahokia and the Hinterlands: Middle Mississippian Cultures of the Midwest.* Urbana and Chicago: University of Illinois Press, 1991.

Epp, Henry, ed. "Henry Kelsey's Journals." In Henry T. Epp, ed., *Three Hundred Prairie Years: Henry Kelsey's "Inland Country of Good Report,"* 196–235. Regina: Canadian Plains Research Center, 1993.

———, ed. *Three Hundred Prairie Years: Henry Kelsey's "Inland Country of Good Report."* Regina: Canadian Plains Research Center, 1993.

Epp, Henry T., and Ian Dyck. *Tracking Ancient Hunters: Prehistoric Archaeology in Saskatchewan.* Regina: Saskatchewan Archaeological Society, 1983.

Ewers, John C. *The Blackfeet: Raiders on the Northwestern Plains.* Norman: University of Oklahoma Press, 1958.

———. *Blackfeet Indians: Ethnological Report on the Blackfeet and Gros Ventre Tribes of Indians.* New York: Garland, 1974.

———. "The Emergence of the Plains Indian as the Symbol of the North American Indian." *Annual Report of the Smithsonian Institution,* 1964 (1965): 531–44.

———. *The Horse in Blackfeet Indian Culture: With Comparative Material from Other Western Tribes.* Washington, D.C.: Smithsonian Institution Press, 1955.

———. *Indian Life on the Upper Missouri.* Norman: University of Oklahoma Press, 1968.

———. "The Indian Trade of the Upper Missouri before Lewis and Clark: An Interpretation." *Missouri Historical Society Bulletin* 10 (4) (July 1954): 429–46.

———, ed. *Crow Indian Medicine Bundles.* Contributions from the Museum of the American Indian, Heye Foundation 17 (1960). New York: Museum of the American Indian, 1960.

Fahey, John. *The Flathead Indians.* Norman: University of Oklahoma Press, 1974.

Fairfield, David J. "Chesterfield House and the Bow River Expedition." M.A. thesis, University of Alberta, 1970.

Fitzgerald, R. D., and A. W. Bailey. "Control of Aspen Regrowth by Grazing with Cattle." *Journal of Range Management* 37 (1984): 156–58.

Fitzgerald, R. D., R. J. Hudson, and A. W. Bailey. "Grazing Preferences of Cattle in Regenerating Aspen Forest." *Journal of Range Management* 39 (1986): 13–18.

Fleming, R. Harvey. "The Origin of 'Sir Alexander Mackenzie and Company.'" *Canadian Historical Review* 9 (1928): 137–55.

Flinn, Michael, ed. *Scottish Population History: From the Seventeenth Century to the 1930s.* Cambridge: Cambridge University Press, 1977.

Flores, Dan. "Bison Ecology and Bison Diplomacy: The Southern Plains from 1800 to 1850." *Journal of American History* 78 (1991): 465–85.

———. "Place: An Argument for Bioregional History." *Environmental History Review* 18 (Winter 1994): 1–18.

Forbis, Richard G. *Cluny: An Ancient Fortified Village in Alberta.* Occasional Paper No. 4. Calgary: Department of Archaeology of the University of Calgary, 1977.

———. *The Old Women's Buffalo Jump, Alberta.* Bulletin No. 180. Ottawa: National Museum of Canada, 1960.

Foster, John E. "Wintering, the Outsider Adult Male and the Ethnogenesis of the Western Plains Métis." *Prairie Forum* 19 (1994): 1–13.

Foster, Michael K. "Language and the Culture History of North America." In Ives Goddard, ed., *Languages,* 64–110. Vol. 17 of William G. Sturtevant, gen. ed., *Handbook of North American Indians.* Washington, D.C.: Smithsonian Institution Press, 1996.

Fowler, Loretta. *Shared Symbols, Contested Meanings: Gros Ventre Culture and History, 1778–1984.* Ithaca, N.Y.: Cornell University Press 1987.

Frison, George C. *Prehistoric Hunters of the High Plains.* New York: Academic Press, 1978.

———. "The Role of Buffalo Procurement in Post-Altithermal Populations on the Northwestern Plains." *University of Michigan Museum of Anthropology Anthropological Papers* 46 (1972): 11–20.

Garry, Nicholas. "Diary of Nicholas Garry, Deputy-Governor of the Hudson's Bay Company from 1822–1835." *Transactions of the Royal Society of Canada,* series 2, vol. 6 (1900), section 2: 73–204.

Gates, C. C., T. Chowns, and H. Reynolds. "Wood Buffalo at the Crossroads." In J. Foster, D. Harrison, and I. S. MacLaren, eds., *Buffalo,* 139–65. Edmonton: University of Alberta Press, 1992.

Getty, Ronald M., and Knut R. Fladmark, eds. *Historical Archaeology in Northwestern North America.* Calgary: University of Calgary Archaeology Association, 1973.

Gibson, Jim. Untitled manuscript discussing an Atsina map drawn in 1801. Deposited in the Hudson's Bay Company Archives, PP 1988-14.

Given, Brian J. *A Most Pernicious Thing: Gun Trading and Native Warfare in the Early Contact Period.* Ottawa: Carleton University Press, 1994.

Glover, Richard, ed. *David Thompson's Narrative: 1784–1812.* Toronto: Champlain Society, 1962.

Gordon, Bryan H. C. *Of Men and Herds in Canadian Plains Prehistory.* Mercury Series Paper No. 84. Ottawa: National Museums of Man, Archeological Survey of Canada, 1979.

Gough, Barry. *First Across the Continent: Sir Alexander Mackenzie.* Norman: University of Oklahoma Press, 1997.

———, ed. *The Journal of Alexander Henry the Younger: 1799–1814.* 2 vols. Toronto: Champlain Society, 1992.

Graspointer, Andreas. *Archaeology and Ethno-history of the Milk River in Southern Alberta.* Calgary: Western Publishers, 1980.

Gregg, Michael L. "Archaeological Complexes of the Northeastern Plains and Prairie-Woodland Border, A.D. 500–1500." In Karl H. Schlesier, ed., *Plains Indians, A.D. 500–1500: The Archaeological Past of Historic Groups,* 71–95. Norman: University of Oklahoma Press, 1994.

———. *An Overview of the Prehistory of Western and Central North Dakota.* Cultural Resources Series No. 1. Billings, Mont.: Bureau of Land Management, 1985.

Greiser, Sally T. "Late Prehistoric Cultures on the Montana Plains." In Karl H. Schlesier, ed., *Plains Indians, A.D. 500–1500: The Archaeological Past of Historic Groups,* 34–55. Norman: University of Oklahoma Press, 1994.

Grinnell, George Bird. *Blackfeet Indian Stories.* New York: Charles Scribner's Sons, 1913.

———. *Blackfoot Lodge Tales: The Story of a Prairie People.* Lincoln: University of Nebraska Press, 1921 [1892].

———. "Coup and Scalping among the Plains Indians." *American Anthropologist* 12 (1910): 296–310.

Grove, Richard. *Ecology, Climate and Empire: Colonialism and Global Environmental History, 1400–1940.* Cambridge: White Horse Press, 1997.

Haas, Jonathan, ed. *The Anthropology of War.* Cambridge: Cambridge University Press, 1990.

Haas, Mary R. "Is Kutenai Related to Algonkian?" *Canadian Journal of Linguistics* 10 (1965): 77–92.

Habgood, Thelma. "Petroglyphs and Pictographs in Alberta." *Archaeological Society of Alberta Newsletter* 13/14 (Summer 1967): 1–40.

Haines, Francis. *Horses in America.* New York: Crowell, 1971.

———. "The Northward Spread of Horses among the Plains Indians." *American Anthropologist* 40 (1938): 429–37.

———. "Where Did the Plains Indians Get Their Horses?" *American Anthropologist* 40 (1938): 112–17.

Hale, Horatio. "Report on the Blackfoot Tribes." *Report of the British Association for the Advancement of Science* 55 (1886): 696–708.

Hall, Robert L. "Cahokia Identity and Interaction Models of Cahokia Mississippian." In Thomas E. Emerson and R. Barry Lewis, eds., *Cahokia and the Hinterlands: Middle Mississippian Cultures of the Midwest,* 3–34. Urbana and Chicago: University of Illinois Press, 1991.

Hamilton, Scott. "Competition and Warfare: Functional versus Historical Explanations." *Canadian Journal of Native Studies* 5 (1985): 93–114.

———. "Western Canadian Fur Trade History and Archaeology: The Illumination of the Invisible in Fur Trade Society." *Saskatchewan Archaeology* 11/12 (1990–91): 3–24.

Hanson, Charles E., Jr. "The Mexican Traders." *Museum of the Fur Trade Quarterly* 6 (3) (1970): 2–6.

Hanson, James A. "Spain on the Plains." *Nebraska History* 74 (1) (1993): 2–21.

Hanson, Jeffery. "Bison Ecology in the Northern Plains and a Reconstruction of Bison Patterns for the North Dakota Region." *Plains Anthropologist* 29 (1984): 93–113.

Harris, R. Cole, ed. *Historical Atlas of Canada.* Vol. 1, *From the Beginning to 1800.* Toronto: University of Toronto Press, 1987.

Helm, June. " 'Always with Them Either a Feast or a Famine': Living Off the Land with Chipewyan Indians, 1791–1792." *Arctic Anthropology* 30 (2) (1993): 46–60.

———. "Bilaterality in the Socio-Territorial Organization of the Arctic Drainage Dene." *Ethnology* 4 (1965): 361–85.

———. *The Lynx Point People: The Dynamics of a Northern Athapaskan Band.* Anthropological Series 53. Bulletin 176. Ottawa: National Museum of Canada, 1961.

———, ed. *Essays on the Problem of Tribe: Proceedings of the 1967 Annual Spring Meeting of the American Ethnological Society.* Seattle: University of Washington Press, 1968.

———, ed. *Subarctic.* Vol. 6 of William G. Sturtevant, gen. ed., *Handbook of North American Indians.* Washington, D.C.: Smithsonian Institution Press, 1981.

Hickerson, Harold. *The Chippewa and Their Neighbours: A Study in Ethnohistory.* Rev. ed. Prospect Heights, Ill.: Waveland Press, 1988.

Hildebrand, David V., and Geoffrey A. J. Scott. "Relationships between Moisture Deficiency and Amount of Tree Cover on the Pre-agricultural Canadian Prairies." *Prairie Forum* 12 (1987): 203–16.

Hodge, Frederick Webb. *Handbook of American Indians North of Mexico.* 2 vols. Grosse Pointe, Mich.: Scholarly Press, 1912 [rept. 1968].

Hoebel, E. Adamson. Review of Oscar Lewis, *Effects of White Contact upon Blackfoot Culture. American Anthropologist* 45 (1943): 464–65.

Hopkins, Donald R. *Princes and Peasants: Smallpox in History.* Chicago: University of Chicago Press, 1983.

Hopwood, Victor G., ed. *David Thompson: Travels in Western North America, 1784–1812.* Toronto: Macmillan, 1971.

Horton, Percy Russell. "Some Effects of Defoliation on Plains Rough Fescue (*Festuca hallii* [Vasey] Piper) in Central Alberta." Ph.D. dissertation, University of Alberta, 1991.

Howell, Signe, and Roy Willis, eds. *Societies at Peace: Anthropological Perspectives.* London: Routledge, 1989.

Hoxie, Frederick E. *Parading through History: The Making of the Crow Nation in America, 1805–1935.* Cambridge: Cambridge University Press, 1995.

Hsu, Francis L. K. "Rethinking the Concept 'Primitive.'" *Current Anthropology* 5 (1964): 169–78.

Hudson, R. J., and S. Frank. "Foraging Ecology of Bison in Aspen Boreal Habitats." *Journal of Range Management* 40 (1987): 71–75.

Hurt, R. Douglas. *Indian Agriculture in America: Prehistory to the Present.* Lawrence: University Press of Kansas, 1987.

Innis, Harold A. *The Fur Trade in Canada: An Introduction to Canadian Economic History.* Toronto: University of Toronto Press, 1956 [1930].

———. "The North West Company." *Canadian Historical Review* 8 (1927): 308–21.

Ives, John W. *A Theory of Northern Athapaskan Prehistory.* Boulder, Colo., and Calgary: Westview Press, 1990.

Jackson, John C. "Brandon House and the Mandan Connection." *North Dakota History* 49 (1982): 11–19.

Jenness, Diamond. *The Sarcee Indians of Alberta.* Bulletin No. 90. Ottawa: Department of Mines and Resources and National Museum of Canada, 1938.

Jennings, Francis. "A Growing Partnership: Historians, Anthropologists and American Indian History." *Ethnohistory* 29 (1982): 21–34.

Johnson, Alice M., ed. *Saskatchewan River Journals and Correspondence: 1795–1802.* London: Hudson's Bay Record Society, 1967.

Johnson, Ann M. "Problem of Crow Pottery." In Leslie B. Davis, ed., *Symposium on the Crow-Hidatsa Separations.* Special issue of *Archaeology in Montana* 20 (3) (September–December 1979): 17–29.

Johnston, A., and S. Smoliak. "Reclaiming Brushland in Southwestern Alberta." *Journal of Range Management* 21 (1968): 404–6.

Judd, Carol M., and Arthur J. Ray, eds. *Old Trails and New Directions: Papers of the Third North American Fur Trade Conference.* Toronto: University of Toronto Press, 1980.

Kavanagh, Thomas W. *Comanche Political History: An Ethnohistorical Perspective 1706–1875.* Lincoln: University of Nebraska Press, 1996.

Keeley, Lawrence H. *War before Civilization: The Myth of the Peaceful Savage.* Oxford: Oxford University Press, 1997.

Kehoe, Alice B. "The Function of Ceremonial Sexual Intercourse among the Northern Plains Indians." *Plains Anthropologist* 15 (1970): 99–103.

———. "How the Ancient Peigans Lived." *Research in Economic Anthropology* 14 (1993): 87–105.

Kehoe, Thomas F. *The Gull Lake Site: A Prehistoric Bison Drive Site in Southwestern Saskatchewan.* Milwaukee: Milwaukee Public Museum, 1973.

———. "Paleo-Indian Drives." In Leslie B. Davis and Michael Wilson, eds., *Bison Procurement and Utilization: A Symposium.* Memoir 14, *Plains Anthropologist* 23 (pt. 2) (1978): 79–83.

Kehoe, Thomas F., and Alice B. Kehoe. "The Identification of the Fall or Rapid Indians." *Plains Anthropologist* 19 (65) (1974): 231–32.

———. "Note to 'The Identification of the Fall or Rapid Indians.'" *Plains Anthropologist* 19 (66) (1974): 302.

Keyser, J. D. "The Plains Indian War Complex and the Rock Art of Writing-on-Stone, Alberta, Canada." *Journal of Field Archaeology* 6 (1979): 41–48.

———. "A Shoshonean Origin for the Plains Shield Bearing Warrior Motif." *Plains Anthropologist* 20 (1975): 207–16.

———. "Writing-on-Stone: Rock Art on the Northwestern Plains." *Canadian Journal of Archaeology* 1 (1977): 15–80.

Kidd, Kenneth E. *Blackfoot Ethnology.* Manuscript Series No. 8. Edmonton: Archaeological Survey of Alberta, 1986.

Klimko, Olga. "New Perspectives on Avonlea: A View from the Saskatchewan Forest." In David Burley, ed., *Contributions to Plains Prehistory*, 64–81. Occasional Paper No. 26, Alberta Culture. Edmonton: Archaeological Survey of Alberta, 1985.

Kozlowski, T. T., ed. *Fire and Ecosystems.* New York: Academic Press, 1974.

Kroeber, Alfred L. *The Arapaho. Bulletin of the American Museum of Natural History* 18 (1902).

———. "Ethnology of the Gros Ventre." *American Museum of Natural History Anthropological Papers* 1 (1908): 145–281.

Küchler, A. W. *Potential Natural Vegetation of the Coterminous United States.* New York: American Geographical Society, 1964.

Lamb, W. K., ed. *Sixteen Years in the Indian Country: Journal of Daniel Williams Harmon, 1800–1816.* Toronto: Macmillan, 1957.

Larson, Floyd. "The Role of Bison in Maintaining the Short Grass Plains." *Ecology* 21 (2) (1940): 113–21.

Le Raye, Charles. "The Journal of Charles Le Raye." *South Dakota Historical Collections* (1908): 150–80.

Lewis, Henry T. "Fire Technology and Resource Management." In Nancy M. Williams and Eugene S. Hunn, eds., *Resource Managers: North American and Australian Hunter-Gatherers*, 45–67. Boulder, Colo.: Westview Press, 1982.

———. "Maskuta: The Ecology of Indian Fires in Northern Alberta." *Western Canadian Journal of Anthropology* 7 (1977): 15–52.

———. *A Time for Burning: Traditional Uses of Fire in the Western Canadian Boreal Forest.* Edmonton: Boreal Institute for Northern Studies, 1982.

Lewis, Oscar. *The Effects of White Contact upon Blackfoot Culture, with Special Reference to the Role of the Fur Trade.* New York: J. J. Augustin, 1942.

Longley, Richmond W. *The Climate of the Prairie Provinces.* Climatological Studies No. 13. Toronto: Environment Canada, 1972.

———. "The Frequency of Chinooks in Alberta." *Albertan Geographer* 3 (1966–67): 20–22.

Looman, Jan. *Prairie Grasses Identified and Described by Vegetative Characters.* Ottawa: Canadian Government Publishing Centre, 1982.

Loscheider, Mavis A. "Use of Fire in Interethnic and Intraethnic Relations on the Northern Plains." *Western Canadian Journal of Anthropology* 7 (4) (1977): 82–96.

Lowie, Robert. "The Assiniboine." *American Museum of Natural History Anthropological Papers* 4 (1910): 1–270.

———. *Indians of the Plains.* New York: McGraw Hill, 1954.

———. "The Northern Shoshone." *American Museum of Natural History Anthropological Papers* 2 (1909): 165–306.

Macleod, J. E. A. "Piegan Post and the Blackfoot Trade." *Canadian Historical Review* 24 (1943): 273–79.

Madsen, David B. "Dating Paiute-Shoshoni Expansion in the Great Basin." *American Antiquity* 40 (1975): 82–86.

Madsen, David B., and David Rhode. "Where Are We?" In David B. Madsen and David Rhode, eds., *Across the West: Human Population Movement and the Expansion of the Numa,* 214–19. Salt Lake City: University of Utah Press, 1994.

———, eds. *Across the West: Human Population Movement and the Expansion of the Numa.* Salt Lake City: University of Utah Press, 1994.

Magne, Martin. "Distributions of Native Groups in Western Canada. A.D. 1700 to A.D. 1850." In *Archaeology in Alberta,* 220–32. Occasional Paper No. 31, Alberta Culture. Edmonton: Archaeological Survey of Alberta, 1986.

Magne, Martin, and M. Klassen. "A Multivariate Study of Rock Art Anthropomorphs at Writing-on-Stone, Southern Alberta." *American Antiquity* 56 (1991): 389–418.

Malainey, Mary E. "The Gros Ventre/Fall Indians in Historical and Archaeological Interpretation." Unpublished paper in the author's possession.

Malainey, Mary E., and Barbara L. Sherriff. "Adjusting Our Perceptions: Historical and Archaeologist Evidence of Winter on the Plains of Western Canada." *Plains Anthropologist* (1996): 333–57.

Mandelbaum, David G. *The Plains Cree: An Ethnographic, Historical and Comparative Study.* Regina: Canadian Plains Research Center, 1979.

Martin, Calvin. *Keepers of the Game: Indian-Animal Relationships and the Fur Trade.* Berkeley: University of California Press, 1978.

McCauley, Clark. "Conference Overview." In Jonathan Haas, ed., *The Anthropology of War,* 1–25. Cambridge: Cambridge University Press, 1990.

McConnell, R. G. "Report on the Cypress Hills Wood Mountain and Adjacent Country." Geological Survey of Canada, *Annual Report,* 1885, part C: 1–85.

McDonald of Garth, John. "Autobiographical Notes, 1791–1816." In *Les Bourgeois de la Compagnie du Nord-Ouest.* Vol. 2. Edited by L. R. Masson. Quebec: A. Cote, etc., 1890.

McGinnis, Anthony. *Counting Coup and Cutting Horses: Intertribal Warfare on the Northern Plains 1738–1889.* Evergreen, Colo.: Cordillera Press, 1990.

McIlwraith, Thomas F. "The Progress of Anthropology in Canada." *Canadian Historical Review* 11 (1930): 132–50.

McNeill, William H. *Plagues and Peoples.* Garden City, N.Y.: Anchor Books, 1976.

Meyer, David. "People before Kelsey: An Overview of Cultural Developments." In Henry T. Epp, ed., *Three Hundred Prairie Years: Henry Kelsey's "Inland Country of Good Report,"* 54–73. Regina: Canadian Plains Research Center, 1993.

Meyer, David, and Henry T. Epp. "North-South Interaction in the Late Prehistory of Central Saskatchewan." *Plains Anthropologist* 35 (132) (1990): 321–42.

Meyer, David, and Scott Hamilton. "Neighbors to the North: Peoples of the Boreal Forest." In Karl H. Schlesier, ed., *Plains Indians, A.D. 500–1500: The Archaeological Past of Historic Groups,* 96–127. Norman: University of Oklahoma Press, 1994.

Meyer, David, and Paul Thistle. "Saskatchewan River Rendezvous Centers and Trading Posts: Continuity in a Cree Social Geography." *Ethnohistory* 42 (1995): 403–44.

Milloy, John S. *The Plains Cree: Trade, Diplomacy and War 1790 to 1870.* Winnipeg: University of Manitoba Press, 1988.

Moodie, D. W. "The Trading Post Settlement of the Canadian Northwest, 1774–1821." *Journal of Historical Geography* 13 (1987): 360–74.

Moodie, D. Wayne, A. J. W. Catchpole, and Kerry Abel. "Northern Athapaskan Oral Traditions and the White River Volcano." *Ethnohistory* 39 (1992): 148–71.

Moodie, D. W., and Barry Kaye. "The Ac Ko Mok Ki Map." *Beaver* 307 (4) (1977): 4–15.

Moodie, D. W., and A. J. Ray. "Buffalo Migrations in the Canadian Plains." *Plains Anthropologist* 21 (1976): 45–52.

Mooney, James. *The Ghost-Dance Religion and the Sioux Outbreak of 1890.* United States Bureau of Ethnology Annual Report to the Secretary of the Smithsonian Institution 14 (1896): part 2.

Moore, John H. "Putting Anthropology Back Together Again: The Ethnogenetic Critique of Cladistic Theory." *American Anthropologist* 96 (1994): 925–48.

Morgan, Lawrence R. "Kootenay-Salishan Linguistic Comparison: A Preliminary Study." M.A. thesis, University of British Columbia, 1980.

Morgan, R. Grace. "Beaver Ecology/Beaver Mythology." Ph.D. dissertation, University of Alberta, 1991.

———. "Bison Movement Patterns on the Canadian Plains: An Ecological Analysis." *Plains Anthropologist* 25 (1980): 143–60.

———. *An Ecological Study of the Northern Plains as Seen through the Garrat Site.* Occasional Paper in Anthropology No. 1. Regina: University of Regina, Department of Anthropology, 1979.

Morton, Arthur S. *A History of the Canadian West to 1870–71.* Toronto: University of Toronto Press, 1973 [1939].

———, ed. *The Journal of Duncan M'Gillivray of the North West Company at Fort George on the Saskatchewan, 1794–5.* Toronto: Macmillan, 1929.

Moss, E. H. "The Vegetation of Alberta." *Botanical Review* 21 (1955): 494–62.

Moulton, Gary E., ed. *The Journals of the Lewis and Clark Expedition.* 11 vols. Lincoln: University of Nebraska Press, 1983–97.

Mulloy, William T. *The Hagen Site: A Prehistoric Village on the Lower Yellowstone.* Publications in the Social Sciences No. 1. Missoula: University of Montana, 1942.

Murphy, Robert F., and Yolanda Murphy. "Northern Shoshone and Bannok." In Warren L. D'Azevedo, ed., *Handbook of North American Indians,* vol. 11, *Great Basin,* 284–307. Washington, D.C.: Smithsonian Institution Press, 1986.

Nelson, J. G. *The Last Refuge.* Montreal: Harvest House, 1973.

Nelson, J. G., and R. E. England. "Some Comments on the Causes and Effects of Fire in the Northern Grasslands Area of Canada and the Nearby United States, ca. 1750–1900." *Canadian Geographer* 15 (1971): 295–306.

Newcomb, W. W., Jr. "A Re-examination of the Causes of Plains Warfare." *American Anthropologist* 52 (1950): 317–30.

Nicholson, Beverly A. "Ceramic Affiliations and the Case for Incipient Horticulture in Southwestern Manitoba." *Canadian Journal of Archaeology* 14 (1990): 33–59.

———. "Interactive Dynamics of Intensive Horticultural Groups Coalescing in South-Central Manitoba during the Late Prehistoric Period—The Vickers Focus." *North American Archaeologist* 15 (1994): 103–27.

———. "Orientation of Burials and Patterning in the Selection of Sites for Late Prehistoric Burial Mounds in South-Central Manitoba." *Plains Anthropologist* 39 (148) (1994): 161–71.

Nicks, John. "Orkneymen in the HBC 1780–1821." In Carol M. Judd and Arthur J. Ray, eds., *Old Trails and New Directions: Papers of the Third North American Fur Trade Conference,* 102–26. Toronto: University of Toronto Press, 1980.

Northwest Power Planning Council. *Compilation of Information on Salmon and Steelhead Losses in the Columbia River Basin.* Portland, Ore.: Northwest Power Planning Council, 1986.

Oliver, Symmes C. *Ecology and Cultural Continuity as Contributing Factors in the Social Organization of the Plains Indian.* Berkeley and Los Angeles: University of California Press, 1962.

Osborn, Alan J. "Ecological Aspects of Equestrian Adaptations in North America." *American Anthropologist* 85 (1983): 563–91.

Owen, Roger C. "The Patrilocal Band: A Linguistically and Culturally Hybrid Social Unit." *American Anthropologist* 67 (1965): 675–90.

Payne, Gene F. *Vegetative Rangeland Types in Montana.* Bulletin 671. Bozeman: Montana Agricultural Experimental Station, Montana State University, 1973.

Peden, D. G., G. M. Van Dyne, R. W. Rice, and R. M. Hansen. "The Trophic Ecology of *Bison bison* L. on Shortgrass Plains." *Journal of Applied Ecology* 11 (1974): 489–97.

Peers, Laura. *The Ojibwa of Western Canada: 1780 to 1870.* Winnipeg: University of Manitoba Press, 1994.

Peterson, Jacqueline, and Jennifer S. H. Brown, eds. *The New Peoples: Being and Becoming Métis in North America.* Winnipeg: University of Manitoba Press, 1985.

Pettipas, Leo. "The Hidatsas in Manitoba." *Manitoba Archaeological Journal* 2 (2) (1992): 11–20.

Phillips, Paul Chrisler. *The Fur Trade.* Norman: University of Oklahoma Press, 1961.

Pijoan, Michel. "The Herds of Oñate: A Speculation." *El Palacio* 81 (3) (1975): 8–17.

Pyne, Stephen J. *Fire in America: A Cultural History of Wildland and Rural Fire.* Princeton, N.J.: Princeton University Press, 1982.

Raby, S. "Prairie Fires in Saskatchewan Grassland." *Saskatchewan History* 19 (1966): 81–99.

Ray, Arthur J. *Indians in the Fur Trade: Their Role as Hunters, Trappers and Middlemen in the Lands Southwest of Hudson Bay 1660–1870.* Toronto: University of Toronto Press, 1974.

———. "The Northern Great Plains, Pantry of the Northwestern Fur Trade, 1774–1855." *Prairie Forum* 9 (1984): 263–80.

Ray, Arthur J., and Donald B. Freeman. *"Give Us Good Measure": An Economic Analysis of Relations between the Indians and the Hudson's Bay Company before 1763.* Toronto: University of Toronto Press, 1978.

Reeves, Brian O. K. "Bison Killing in the Southwestern Rockies." In Leslie B. Davis and Michael Wilson, eds., *Bison Procurement and Utilization: A Symposium,* 63–78. Memoir 14, *Plains Anthropologist* 23 (pt. 2) (1978).

———. "Communal Bison Hunters." In Leslie B. Davis and B. O. K. Reeves, eds., *Hunters of the Recent Past,* 168–94. London: Unwin Hyman, 1990.

———. *Culture Change in the Northern Plains 1000 B.C.–A.D. 1000.* Occasional Paper No. 20. Edmonton: Archaeological Survey of Alberta, 1983.

Rich, E. E. *The Fur Trade and the Northwest to 1857.* Toronto: McClelland and Stewart, 1967.

———. *The History of the Hudson's Bay Company 1670–1870.* 3 vols. Toronto: McClelland and Stewart, 1960.

———. "Trade Habits and Economic Motivation among the Indians of North America." *Canadian Journal of Economics and Political Science* 26 (1960): 35–53.

———, ed. *Cumberland House Journals and Inland Journals 1775–1779.* London: Hudson's Bay Record Society, 1951.

———, ed. *Cumberland House Journals and Inland Journals 1779–1782.* London: Hudson's Bay Record Society, 1952.

———, ed. *James Isham's Observations on Hudson's Bay, 1743, and Notes and Observations on a Book Entitled "A Voyage to Hudson's Bay in the Dobbs Galley, 1749."* Toronto: Champlain Society, 1949.

Richmond, R. J., R. J. Hudson, and R. J. Christopherson. "Comparison of Forage Intake and Digestibility of Bison, Yak, and Cattle." *Annual Feeders' Day Report* (8 June 1976): 49–50.

Rinn, Dennis Lloyd. "The Acquisition, Diffusion and Distribution of the European Horse among the Blackfoot Tribes in Western Canada." M.A. thesis, University of Manitoba, 1975.

Roe, Frank Gilbert. *The Indian and the Horse.* Norman: University of Oklahoma Press, 1955.

———. *The North American Buffalo: A Critical Study of the Species in Its Wild State.* Toronto: University of Toronto Press, 1970.

Rowe, J. Stan. "Lightning Fires in Saskatchewan Grassland." *Canadian Field Naturalist* 83 (1969): 317–24.

Rowe, J. Stan, and Robert T. Coupland. "Vegetation of the Canadian Plains." *Prairie Forum* 9 (1984): 231–48.

Russell, Dale R. *Eighteenth-Century Western Cree and Their Neighbours.* Archaeological Survey of Canada, Mercury Series Paper 143. Ottawa: Canadian Museum of Civilization, 1991.

———. "Native Groups in the Saskatoon Area in the 1700s and 1800s." In Urve Linnamae and Tim E. H. Jones, eds., *Out of the Past: Sites, Digs and Artifacts in the Saskatchewan Area,* 131–83. Saskatoon: Saskatoon Archaeological Society, 1988.

———. "The Puzzle of Henry Kelsey and His Journey to the West." In Henry T. Epp, ed., *Three Hundred Prairie Years: Henry Kelsey's "Inland Country of Good Report,"* 74–88. Regina: Canadian Plains Research Center, 1993.

Russell, Emily W. B. "Indian-Set Fires in the Forests of the Northeastern United States." *Ecology* 64 (1983): 78–88.

Sahlins, Marshall. *Islands of History.* Chicago: University of Chicago Press, 1985.

———. "The Return of the Event, Again: With Reflections on the Beginnings of the Great Fijian War of 1843 to 1855 between the Kingdoms of Bau and Rewa." In Aletta Biersack, ed., *Clio in Oceania: Toward a Historical Anthropology,* 37–99. Washington, D.C.: Smithsonian Institution Press, 1991.

———. *Tribesmen.* Englewood Cliffs, N.J.: Prentice-Hall, 1966.

Sauer, C. O. "Grassland Climax, Fire, and Man." *Journal of Range Management* 3 (1950): 16–21.

Savishinsky, Joel S. "Mobility as an Aspect of Stress in an Arctic Community." *American Anthropologist* 73 (1971): 604–18.

Schaeffer, Claude E. "Plains Kutenai: An Ethnological Evaluation." *Alberta History* 30 (4) (1982): 1–9.

Schlesier, Karl H. "Commentary: A History of Ethnic Groups in the Great Plains A.D. 150–1550." In Karl H. Schlesier, ed., *Plains Indians, A.D. 500–1500: The Archaeological Past of Historic Groups,* 308–81. Norman: University of Oklahoma Press, 1994.

———, ed. *Plains Indians, A.D. 500–1500: The Archaeological Past of Historic Groups.* Norman: University of Oklahoma Press, 1994.

Secoy, Frank Raymond. *Changing Military Patterns of the Great Plains.* Locust Valley, N.Y.: J. J. Augustin, 1953.

Shannon, Raymond, and Daniel J. Smith. "Observations of the Precipitation Regime of the Cypress Hills Area, Alberta and Saskatchewan." *Albertan Geographer* 23 (1987): 33–44.

Sharrock, Susan R. "Crees, Cree-Assiniboines, and Assiniboines: Interethnic Social Organization of the Far Northern Plains." *Ethnohistory* 21 (2) (1974): 95–122.

Shimkin, Demitri B. "Comanche-Shoshone Words of Acculturation, 1786–1848." *Journal of the Steward Anthropology Society* 11 (2) (1980): 195–248.

———. "Eastern Shoshone." In Warren L. D'Azevedo, ed., *Handbook of North American Indians,* vol. 11, *Great Basin,* 308–34. Washington, D.C.: Smithsonian Institution Press, 1986.

———. "The Introduction of the Horse." In Warren L. D'Azevedo, ed., *Handbook of North American Indians,* vol. 11, *Great Basin,* 517–24. Washington, D.C.: Smithsonian Institution Press, 1986.

———. "Shoshone-Comanche Origins and Migrations." In *Proceedings of the Sixth Pacific Science Congress IV,* 17–25. Berkeley: University of California Press, 1940.

Siebert, Frank T., Jr. "The Original Home of the Proto-Algonquian People." In *Contributions to Anthropology: Linguistics I (Algonquian),* 13–47. Anthropological Series 78, Bulletin 214. Ottawa: National Museum of Canada, 1967.

Simmons, Marc. *The Last Conquistador: Juan de Oñate and the Settling of the Far Southwest.* Norman: University of Oklahoma Press, 1991.

———. "New Mexico's Smallpox Epidemic of 1780–1781." *New Mexico Historical Review* 41 (1966): 319–26.

Simms, Steven R. "Comments on Bettinger and Baumhoff's Explanation of the 'Numic Spread' in the Great Basin." *American Antiquity* 48 (1983): 825–30.

Smith, James G. E. "The Western Woods Cree: Anthropological Myth and Historical Reality." *American Ethnologist* 14 (1987): 434–48.

Smith, Marian W. "The War Complex of the Plains Indians." *Proceedings of the American Philosophical Society* 78 (1938): 425–64.

Smoliak, S., W. D. Willms, and N. W. Holt. *Management of Prairie Rangeland.* Publication 1589/E. Ottawa: Agriculture Canada, 1990.

Smythe, Terry. "Thematic Study of the Fur Trade in the Canadian West: 1670–1870." Typescript. Ottawa: Historic Sites and Monuments Board of Canada, 1968.

Speck, F. G. "The Family Hunting Band as the Basis of Algonkian Social Organization." *American Anthropologist* 17 (1915): 289–305.

Speth, John D., and Katherine A. Spielmann. "Energy Source, Protein Metabolism, and Hunter-Gatherer Subsistence Strategies." *Journal of Anthropological Archaeology* 2 (1983): 1–31.

Spry, Irene M., ed. *The Papers of the Palliser Expedition, 1857–1860.* Toronto: Champlain Society, 1968.

Stewart, Frank H. "Mandan and Hidatsa Villages in the Eighteenth and Nineteenth Centuries." *Plains Anthropologist* 19 (1974): 287–302.

Stewart, Omer C. "Why Were the Prairies Treeless?" *Southwestern Lore* 20 (1955): 59–64.

Stoltman, James B., ed. *New Perspectives on Cahokia: Views from the Periphery.* Madison, Wis.: Prehistory Press, 1991.

Strong, W. D. "From History to Prehistory in the Northern Great Plains." *Smithsonian Miscellaneous Collections* 100 (1940): 353–94.

Strong, W. L., and K. R. Leggat. *Ecoregions of Alberta.* Edmonton: Alberta Forestry, Lands and Wildlife, 1992.

Suttles, Wayne, and William W. Elmendorf. "Linguistic Evidence for Salish Prehistory." In *Symposium on Language and Culture: Proceedings of the 1962 Annual Meeting of the American Ethnological Society,* 40–52. Seattle: American Ethnological Society, 1963.

Sutton, Mark Q. "Warfare and Expansion: An Ethnohistoric Perspective on the Numic Spread." *Journal of California and Great Basin Anthropology* 8 (1986): 65–82.

Swagerty, William R. "Indian Trade in the Trans-Mississippi West to 1870." In Wilcomb E. Washburn, ed., *Handbook of North American Indians,* vol. 4, *History of Indian-White Relations,* 351–74. Washington, D.C.: Smithsonian Institution Press, 1988.

Tax, Sol. "Primitive Peoples." *Current Anthropology* 1 (1960): 441.

Taylor, J. F. "Sociocultural Effects of Epidemics on the Northern Plains: 1734–1850." *Western Canadian Journal of Anthropology* 7 (4) (1977): 55–81.

Taylor, Robert L., Milton J. Edie, and Charles R. Gritzner. *Montana in Maps.* Bozeman: Montana State University, Big Sky Books, 1974.

Teit, James A. "The Salishan Tribes of the Western Plateaus." *Bureau of American Ethnology Annual Report* 45 (1927–28): 23–396.

Thistle, Paul C. *Indian-European Trade Relations in the Lower Saskatchewan River Region to 1840.* Winnipeg: University of Manitoba, 1986.

Thomas, Alfred B. *The Plains Indians and New Mexico, 1751–1778.* Albuquerque: University of New Mexico Press, 1940.

Thwaites, Reuben Gold, ed. *Early Western Travels, 1784–1897.* 32 vols. Cleveland: A. Clark, 1904–7.

———, ed. *Original Journals of the Lewis and Clark Expedition, 1804–1806.* New York: Antiquarian Press, 1959 [1904].

Tiffany, Joseph. "An Overview of the Middle Missouri Tradition." In Guy E. Gibbon, ed., *Prairie Archaeology,* 87–108. Publications in Anthropology No. 3. Minneapolis: University of Minnesota Press, 1983.

Trigger, Bruce G. "Archaeology and the Ethnographic Present." *Anthropolgica* 23 (1981): 3–18.

———. "Early Native North American Responses to European Contact: Romantic versus Rationalistic Interpretations." *Journal of American History* 77 (1991): 1195–1215.

———. *Natives and Newcomers: Canada's "Heroic Age" Reconsidered.* Kingston and Montreal: McGill–Queen's University Press, 1985.

Turnbull, Colin M. "The Importance of Flux in Two Hunting Societies." In Richard B. Lee and Irven DeVore, eds., *Man the Hunter,* 132–37. Chicago: Aldine, 1968.

Turner, B. L., II, and Karl W. Butzer. "The Columbian Encounter and Land-Use Change." *Environment* 34 (8) (October 1992): 16–45.

Turney-High, Harry Holbert. *Ethnography of the Kutenai.* Memoirs No. 56. Menasha, Wis.: American Anthropological Association, 1941.

———. *Primitive Warfare: Its Practice and Concepts.* Columbia: University of South Carolina Press, 1971 [1949].

Tyrrell, J. B., ed. *David Thompson's Narrative of His Explorations in Western America, 1784–1812.* Toronto: Champlain Society, 1916.

———, ed. *Journals of Samuel Hearne and Philip Turnor.* Toronto: Champlain, 1934.

Umfreville, Edward. *The Present State of Hudson's Bay.* London: Charles Stalker, 1790.

Upham, Steadman. "Nomads of the Desert West: A Shifting Continuum in Prehistory." *Journal of World Prehistory* 8 (1994): 113–67.

Verbicky-Todd, Eleanor. *Communal Buffalo Hunting among the Plains Indians: An Ethnographic and Historical Review.* Occasional Paper No. 24. Edmonton: Archaeological Survey of Alberta, 1984.

Vickers, J. Roderick. *Alberta Plains Prehistory: A Review.* Occasional Paper No. 27. Edmonton: Archaeological Survey of Alberta, 1986.

———. "Cultures of the Northwestern Plains: From the Boreal Forest Edge to Milk River." In Karl H. Schlesier, ed., *Plains Indians,* A.D. *500–1500: The Archaeological Past of Historic Groups,* 3–33. Norman: University of Oklahoma Press, 1994.

———. "Seasonal Round Problems on the Alberta Plains." *Canadian Journal of Archaeology* 15 (1991): 55–72.

Vrooman, C. W., G. D. Chattaway, and Andrew Stewart. *Cattle Ranching in Western Canada.* Publication No. 778. Ottawa: Department of Agriculture, 1946.

Walde, Dale, David Meyer, and Wendy Unfreed. "The Late Period on the Canadian and Adjacent Plains." *Revista de Arqueología Americana* 9 (July–December 1995): 7–66.

Wallace, W. S. "The Pedlars from Quebec." *Canadian Historical Review* 13 (1932): 387–402.

———, ed. *Documents Relating to the North West Company.* Toronto: Champlain Society, 1934.

Watts, F. B. "The Natural Vegetation of the Southern Great Plains of Canada." *Geographical Bulletin* 14 (1960): 25–43.

Weber, David J. *The Spanish Frontier in North America.* New Haven: Yale University Press, 1992.

———. *The Taos Trappers: The Fur Trade in the Far Southwest, 1540–1846.* Norman: University of Oklahoma Press, 1971.

Weist, Katherine M. "An Ethnohistorical Analysis of Crow Political Alliances." *Western Canadian Journal of Anthropology* 7 (4) (1977): 34–54.

White, Richard. *The Middle Ground: Indians, Empires, and Republics in the Great Lakes Region, 1650–1815.* New York: Cambridge University Press, 1991.

Willey, Patrick Soren. *Prehistoric Warfare on the Great Plains: Skeletal Analysis of the Crow Creek Massacre Victims.* New York: Garland, 1990.

Williams, Glyndwr. "The Hudson's Bay Company and the Fur Trade: 1670–1870." *Beaver* 314 (2) (Autumn 1983): 4–86.

———. "The Puzzle of Anthony Henday's Journal, 1754–1755." *Beaver* 309 (3) (1978): 41–56.

———, ed. *Andrew Graham's Observations on Hudson's Bay, 1767–91.* London: Hudson's Bay Record Society, 1969.

Williams, Nancy M., and Eugene S. Hunn, eds. *Resource Managers: North American and Australian Hunter-Gatherers.* Boulder, Colo.: Westview Press, 1982.

Willms, W. D., S. Smoliak, and J. F. Dorman. "Effects of Stocking Rate on a Rough Fescue Grassland Vegetation." *Journal of Range Management* 38 (1985): 220–27.

Wilson, E. F. "Report on the Blackfoot Tribes." *Report of the British Association for the Advancement of Science* 57 (1887): 183–200.

———. "Report on the Sarcee Indians." *Report of the British Association for the Advancement of Science* 58 (1888): 242–55.

Wilson, Michael. "The Early Historic Fauna of Southern Alberta: Some Steps to Interpretation." In Ronald M. Getty and Knut R. Fladmark, eds., *Historical Archaeology in Northwestern North America,* 213–48. Calgary: University of Calgary Archaeology Association, 1973.

Wishart, David J. *The Fur Trade of the American West 1807–1840: A Geographical Synthesis.* Lincoln: University of Nebraska Press, 1979.

Wissler, Clark. "Ethnographical Problems of the Missouri-Saskatchewan Area." *American Anthropologist* 10 (1908): 197–207.

———. "The Influence of the Horse in the Development of Plains Culture." *American Anthropologist* 16 (1914): 1–25.

———. *The Material Culture of the Blackfoot Indians.* American Museum of Natural History Anthropological Papers 5 (1) (1910).

———. *The Social Life of the Blackfoot Indians.* American Museum of Natural History Anthropological Papers 7 (1912).

———. *Societies and Dance Associations of the Blackfoot Indians.* American Museum of Natural History Anthropological Papers 11 (1913).

Wood, W. Raymond, ed. *Papers in Northern Plains Prehistory and Ethnohistory.* Special Publication No. 10. Sioux Falls: South Dakota Archaeological Society, 1986.

Wood, W. Raymond, and Alan S. Downer. "Notes on the Crow-Hidatsa Schism." *Plains Anthropologist,* Memoir 13 (22) (1977): 83–100.

Wood, W. Raymond, and Margot Liberty. *Anthropology on the Great Plains.* Lincoln: University of Nebraska Press, 1980.

Wood, W. Raymond, and Thomas D. Thiessen. *Early Fur Trade on the Northern Plains: Canadian Traders Among the Mandan and Hidatsa Indians, 1738–1818.* Norman: University of Oklahoma Press, 1985.

Wright, Gary A. "The Shoshonean Migration Problem." *Plains Anthropologist* 23 (1978): 113–37.

Wright, Henry A., and Arthur W. Bailey. *Fire Ecology: United States and Southern Canada.* New York: John Wiley and Sons, 1982.

Yerbury, J. C. "Kootenay Linguistics: An Unsolved Mystery." *Idaho Yesterdays* 22 (1) (Spring 1978): 11–15.

Index